THE CHiCKEN WHiSPERER's GUiDE TO KEEPiNG CHiCKENS

QUARRY

THE CHICKEN WHISPERER'S GUIDE TO KEEPING CHICKENS

BEVERLY MASSACHUSETTS

QUARRY BOOKS

Everything You Need to Know ... and Didn't Know You Needed to Know About Backyard and Urban Chickens

ANDY G. SCHNEIDER
and Dr. Brigid McCrea, Ph.D.

© 2011 by Quarry Books

First published in the United States of America in 2011 by Quarry Books, a member of Quayside Publishing Group 100 Cummings Center Suite 406-L Beverly, Massachusetts 01915-6101 Telephone: (978) 282-9590 Fax: (978) 283-2742 www.quarrybooks.com

The Chicken Whisperer's Guide to Keeping Chickens contains a variety of tips and recommendations. While caution was taken to give safe recommendations, it is impossible to predict the outcome of each suggestion. Neither Andy Schneider, Dr. Brigid McCrea, nor the publisher, Quayside Publishing Group, accepts liability for any mental, financial, or physical harm that arises from following the advice or techniques or using the procedures in this book. Readers should use personal judgment when applying the recommendations of this text.

ISBN-13: 978-1-59253-728-0
ISBN-10: 1-59253-728-6

Digital edition published in 2011
eISBN-13: 978-1-61058-142-4

Library of Congress Cataloging-in-Publication Data available

Design: carol holtz | holtz design
Cover Images: iStockphoto, main image; inset, middle; & flap; Courtesy of The Chicken Whisperer, inset, left; Rick Bennett, inset, right
Illustrations: Judy Love
Special thanks to Teri Myers, Braided Bower Farm

Printed in Singapore

[Blue Cochin hen]

CONTENTS

Golden Bantam

Bantam Cockerel

POWER TO THE POULTRY:
THE GROWING URBAN CHICKEN MOVEMENT

THERE IS A FAST-GROWING MOVEMENT sweeping across the world. It's all about urban chickens, and thousands of people are joining every year.

Keeping chickens in urban backyards is nothing new and at one point, was quite common. The United States government, for example, encouraged it during the Great Depression, when food was harder to come by and a family had to provide for its members. When small grocery stores started popping up on every corner and began carrying ready-to-roast chickens and clean, white eggs, the need for a small backyard flock to feed the family dissipated.

The new urban chicken movement picked up steam in the mid-1990s, and went full speed ahead by 2002. Websites such as www.backyardchickens.com, www.feathersite.com, and www.urbanchickens.net started to take off.

Hatcheries started shipping out millions of chicks to backyards everywhere. Local urban chicken groups formed around the world, and today have anywhere from 50 to more than 1,300 members. To keep up with demand, feed-and-seed stores have increased store space reserved for poultry supplies, and a variety of books about urban chickens now flood the market. Andy even hosts a radio show, "Backyard Poultry with the Chicken Whisperer," during which he welcomes poultry experts from around the world to share their knowledge.

Who is responsible for the rebirth of the urban chicken movement? In our opinion, it's the people involved in the green movement.

A backyard chicken coop in Los Angeles promotes a healthy lifestyle
with fresher eggs and greener living.

The urban farming movement has grown substantially in recent years.

PEOPLE WHO LIVE A GREENER, more ecofriendly lifestyle eat healthier, for the most part. They tend to buy a large portion of their food locally, and they understand that there's nothing more local than one's own backyard. Many already supply their families with homegrown vegetables from backyard gardens, so why not eggs from chickens, too?

The green movement may have initiated growth of the urban backyard movement, but people keep urban chickens for many reasons. In fact, the people joining the backyard chicken movement are quite the melting pot. Through local meetup groups, friendships form between people with absolutely nothing in common but keeping chickens for pets, eggs, composting, fertilizer, insect control, education, or meat. The movement is not based on economic status, age, gender, race, religion, ethnicity, political views, or any other label. No stereotype fits the people who keep urban chickens. We like to say that we've never met a chicken keeper we didn't like.

THE CHICKEN WHISPERER'S GUIDE TO KEEPING CHICKENS

AN UPHILL BATTLE

The urban chicken movement has faced its share of struggles. Hatcheries, for example, have strained to keep up with the growing interest. Since 2007, many have either been backlogged with orders or unable to fulfill them due to requests for rare breeds such as the French Black Copper Marans that sell out first and quickly. Chicken owners (or those who want to be) also face laws—for example, a coop having to sit a specific distance from nearby occupied dwellings—or challenges such as having to get permission from neighbors to keep backyard poultry (we discuss the laws in detail in chapter 3). In addition, chicken owners struggle to understand disease and how to reduce their chickens' risk of becoming ill. (For more about this, read chapter 10.)

The good news is that many smaller hatcheries have popped up to help meet demand while also providing income to entrepreneurs. Lawmakers are starting to change laws in favor of backyard poultry, thanks in part to work by newly formed groups that aim to educate lawmakers and the public. Plus, there are more opportunities from government agriculture departments and in books, videos, and on the radio to learn about diseases, sicknesses, and ways to reduce risk of illness in chickens.

Hatcheries are backlogged with orders for would-be chicken owners as interest in the urban poultry market has soared.

Marans varieties are in high demand due to their striking colors. The French Marans are increasingly more popular because of their darker, chocolate-colored eggs.

ADVANTAGES TO OWNING CHICKENS

The movement is continuing to grow. Is it a fad? We don't think so. As more people educate themselves about what foods they are eating, and what they are actually putting into their bodies, they are starting to look for options outside the supermarkets. Chickens can benefit suburban and rural landowners in so many ways.

Fresh Eggs

One of the top reasons for keeping backyard chickens is the endless supply of fresh, tasty eggs. More people today want to know where their food comes from and having backyard chickens allows them that luxury. As a chicken owner, you determine what goes into your chickens, which also gives you a pretty clear picture of what comes out.

Grocery stores now stock a variety of eggs with different labels, each brand trying to convince you it is the best to buy. When we first noticed these, images popped into our heads about what those words meant. "Free range," for example, must mean that the chickens are out in a pasture scratching for bugs, eating plants and supplement feed, and living a life unconfined. "Organic" must mean a similar situation but with feed supplement from an organic product. You may have done this casual inferring yourself. We've talked to many people who have come to similar conclusions about these words.

A family can have a constant supply of fresh eggs with just a few chickens.

Unfortunately, these labels can be deceiving, and without doing a little research, you may not understand them fully. If you purchase eggs from a grocery store, use a variety of online resources to help you know what goes into your eggs. Even then, the information may add little comfort without you actually seeing these practices in action—something you can obviously do with your own flock.

As with any farm, there are good and bad managers; the best you can do is ensure that the food you buy came from someone you have personally met and whose farm you have visited.

In 2007, *Mother Earth News* compared eggs sold at the supermarket, many of which are raised in small battery cages, with those from various breeds of chickens raised on various types of pasture. (For specific results of this study, check out the sidebar, "Cage-Raised versus Backyard Poultry.") According to this study, hens in the latter group laid more nutritionally sound eggs than those sold commercially. In most situations, chickens raised in the backyard setting live plush lives. Owners do their best to protect their flock and at the same time, provide opportunities for chickens to behave as they naturally would in a pasture. For example, most backyard chickens have ample room to move freely, take dust baths, scratch for seeds and bugs, and so on. For this reason, while we don't have a study to prove it yet, we believe most backyard poultry hens lay eggs that are healthier than store-bought eggs laid by chickens raised in cages.

Great Waste Disposers

Did you know that chickens can eat much of your kitchen leftovers? In fact, they can drastically reduce the amount of garbage you set out on the curb each week if your household frequently has leftovers that would otherwise spoil. While some foods are better than others for these birds, chickens can eat many things including garden waste. Just make sure to remove uneaten leftovers from the pen so mold, decay, and bacteria do not make the birds sick, or attract unwanted rodents or predators.

Cage-Raised versus Backyard Poultry

A nonscientific study by *Mother Earth News* (www.motherearthnews.com/eggs) found that eggs from pasture-raised hens contain the following:

- ✦ One-third less cholesterol
- ✦ One-quarter less saturated fat
- ✦ Two-thirds more vitamin A
- ✦ Two times more omega-3 fatty acids
- ✦ Three times more vitamin E
- ✦ Seven times more beta carotene

ANDY'S ANECDOTE

AFTER OUR ANNUAL SUMMER holiday party, I put watermelon rind in the chicken run (the fenced-in area outside the coop for chickens) and the next thing I know, the rind is paper-thin. Often, the chickens won't leave a scrap. If I find a tomato, partially eaten by a worm or bug, I toss it into the run and watch the birds swarm to devour it! After Halloween, when most people toss their pumpkins in the garbage or compost bin, I cut mine up for the chickens.

Your garden and your chicks will have a harmonious relationship, and nothing will go to waste.

It is certainly fun to interact with your chickens in this way. However, if you care about generating a steady supply of eggs, limit the amount of food scraps and other treats in their diet. (We discuss nutrition and well being more in chapter 9.) Chickens that fill up on table scraps may not get the appropriate nutrients they need to produce eggs because they are too full to eat their balanced feed. In addition, chickens may learn to expect these treats—and will let you know through unhappy chattering when they don't get them. They may even hold off eating their balanced feed to have plenty of room for treats when you do offer them. Chickens are smart. Don't discount their willingness to hold out for the good stuff if they know you provide it daily.

Chickens are happy to partake of your kitchen scraps.

Great Fertilizer

It's plain and simple: Chicken poop makes great fertilizer. While chicken waste can be considered hot, which simply means high in nitrogen, it can help many garden plants thrive. Don't place it directly on plants, but rather mix chicken poop into your compost to create healthy soil. There are several ways to do this.

Several months before you want to plant a garden area, house chickens there to eat insects, weed seedlings, and fertilize the garden. Or collect the waste and bedding material from the run, coop, and nest box, and place it in a compost bin or pile where it can age and the materials can decompose to create nutrient-rich soil. A third option is to create a rotation plan during which you move the chickens between your garden and chicken run. Using this method, the chickens fertilize one area of your yard for an entire year. Then, when the chickens move to a new yard location, mix the soil in the first location by tilling or turning it over by hand. (In chapter 4, read about ways to best fertilize your whole yard rather than just a part of it.)

Insect Control

Chickens love to walk around the yard scratching and foraging for bugs. They eat all kinds of insects including but not limited to ants, spiders, ticks, fleas, slugs, roaches, beetles, small snakes, and even small mice. As mentioned above, chickens can work in your garden before, during, and after you plant to keep away pests. Though some of the insects chickens eat may cause them a problem the first time they're eaten, the chickens will learn their lesson quickly and stay away.

Chicken manure provides rich, organic nourishment for fruits, flowers, and vegetables. But keep in mind that your chickens will destroy tender seedlings and bite mature fruits and vegetables if they have garden access.

A boy holds a black rooster while the turkey his brother
is holding flies out of his arms. Their family raises turkeys, show bantams,
and chickens to enter in 4-H competitions.

Education

Many children today have no idea about the origins of the food they eat. They think broccoli comes from the grocery store and the grocery store gets it from a truck. They also think eggs come from Styrofoam cartons and are only white, not brown, or even blue or green. This problem is common with children raised in urban and suburban areas, but the problem also exists in rural areas with children raised off farms. In fact, children raised in rural areas don't necessarily have any more exposure to food production on a farm than urban and suburban kids.

The backyard poultry movement began in urban and suburban areas, but more of the rural population is starting to get back to the basics with chickens, rabbits, goats, and other small animals. Parents all over have told us that because of the distance between food source and plate, their children do not even recognize basic foods in their original forms.

Not only is the process through which a chicken lays an egg educational, but chickens are easy enough to care for that young children (who are old enough) can take on that responsibility. We've found that children take on more responsibility and stay focused longer when caring for a chicken than for a dog, cat, or other more common household pet. Why? Chickens lay eggs, a reward most pets don't provide. Depending on the flock, these can be a beautiful mix of colors, shapes, and sizes, providing a unique daily surprise for the whole family. In addition, children can participate in chicken clubs, such as 4-H, which often encourage and teach children how to enter chickens in local fairs and poultry shows—places with points to be earned, awards to be received, life skills to be learned, and a fun experience for the whole family.

Entertainment

Believe it or not, chickens are very entertaining—maybe even therapeutic. Chickens really do have individual personalities. Some are friendly and come sit on your lap while you pet them. Others love to bask in the sun while they saturate their bodies with dirt. Some others scratch the dirt until they find a prize worm or bug and then run around the yard fending off the rest of the flock—all who want to steal the tasty morsel.

Chickens are unique animals that bring families, neighbors, and communities together just for the sake of talking about chickens. Are you ready to bring them into your home? Read on. Up next is the anatomy of chickens and selecting the breed or breeds that are right for you.

By observing chickens, children can be educated about
the origin of their foods from an early age.

HENOLOGY:
KNOW THY HEN FOR SHE iS GREAT

YOUR CHICKENS ARE FUN and curious family members and friends. If starting your first flock, have faith that your chickens will become your homegrown guide to enjoying your own backyard. Many people with chickens learn to see a common backyard or household space from a chicken-eyed point of view. What does that mean? Well, depending on your flock members' personalities, it can range from learning to relax with Victorian, ladylike style to bounding around the yard with exuberance. Either way, chickens will teach you to check out all this life has to offer.

They will also earn your respect. That's what this chapter is designed for: to help you understand your new feathered friends. They are small and beautiful, yet mighty. Respect them for the physical feat they put forth every day—in the form of that egg you eat for breakfast.

MATING AND REPRODUCTIVE BEHAVIOR

Chickens do not produce eggs year-round. Rather, they focus on seasons and times of the year most likely to yield the highest reproductive success rate. This often corresponds with spring and its lengthening days. Light is the key here: A hen will lay at her maximum production rate with fourteen hours of light. She will still lay eggs with less light, just not as frequently.

What happens to the birds when we move from dark, short winter days to the longer ones of spring and summer? The addition of enough light stimulates their brains to produce hormones aimed at causing their gonads to grow (they had regressed or grown smaller during the short days of winter). Once the gonads reach full size, they begin to produce sperm (for the male) and eggs (for the female), as well as a few hormones of their own.

[Chickens roost proudly showing off their colorful variety of plumage.]

See the Light

Some of you might be thinking that all of this about fourteen hours of light is absolute poppycock because your hens laid well through the winter. Many factors can influence a flock's egg-producing rate. Usually a new flock of hens lays like gangbusters through its first winter—but don't count on that experience every year. Sometimes, your hens may get ambient light from an outdoor light or street lamp. It doesn't take much light to do the trick, but don't be surprised if the lay rate decreases after the first year. Also, if you use heat lamps in your coop during winter, then your hens may lay better because you are giving them additional light.

Hormones play a huge role in a hen's ability to seek out a high-quality male that can both establish and defend a territory with a variety of foods and suitable shelter. In situations with many suitable potential mates, the males may do battle: spurring at each other, beating on each other with their wings, and doing anything to cause the other pain. This determines which male has the stamina and characteristics the female prefers. Males also perform ritual behaviors including the following:

- **Crowing.** This tells the hen much about the male's health and vigor.
- **Tidbitting.** During this behavior, the male makes short clucks while picking up and dropping either a piece of food or a non-food item like grit. Tidbitting stimulates the female to investigate the item about which he is clucking. (At that time, the male may either mate with the female or perform yet another behavior to convince her of his suitability.)
- **Waltzing.** This is when the rooster spreads one wing toward the ground and does a half-circle dance around the hen. He may kick at this one lowered wing and perhaps lean in toward the hen. A receptive hen will permit the male to breed with her.

The hen, when she is ready to mate, will not run from a rooster. She will stand in place, squat a little, and spread out her wings a tiny bit. Some hens also may stomp their feet. (Many hens also perform this behavior when their human caretakers come near or pet them; it is a sign of respect as you are the dominant member of the coop.) During mating, the rooster steps onto the hen's back, treads slightly, and grasps her wings with his feet. To stabilize his actions, he uses his beak to grab the feathers on the back of the hen's neck. Sometimes he pulls out the hen's feathers by mounting and breeding her, but the feathers will grow back. Be sure to check your hens periodically for scratches from the rooster. Keep his toenails and spurs trimmed to prevent injury to the hen.

The mating action itself is very quick. In fact, the vents, or cloacas, of the rooster and hen touch for only a second or two. The female presents the oviduct so the rooster's sperm immediately enters the correct part of the cloaca. The hen stores the sperm in a special section of the oviduct for several days. One mating for an average hen can fertilize just more than a week's worth of eggs.

Do you need your own rooster to get eggs? This is a common question among first-time and long-time chicken owners alike. Think back to your basic biology because this question encompasses more than just the chicken. Do female animals, whether chickens, horses, or humans, need a male of the species present in order ovulate? The answer is no. The frequency at which a female animal ovulates depends on her species. A sexually mature human typically ovulates once a month unless interrupted by a successful pregnancy. A horse has a breeding season, which usually runs from May to August, but the mare ovulates about every twenty-one days until bred. Dogs, whether Great Danes or Chihuahuas, ovulate twice a year until bred.

The biggest difference between chickens and the non-egg laying animals mentioned above is that the chicken does not carry her young inside of her. She continues to ovulate, or lay an egg, every day whether or not she has been successfully bred. The chicken lays eggs, whether the eggs are fertile or not, outside of her body and then incubates those eggs. So she has no idea whether her efforts will succeed because she does not know how many of her eggs are fertile. Fertilization of the ova or yolk takes place inside the hen, and then the egg forms around the potentially fertile yolk. Everything a chick needs to develop successfully is packaged inside the egg before the hen lays it.

In all likelihood, the chickens you get for your home flock are hybrids bred to lay eggs, one a day, for many months in a row, and no, you do not need a rooster. A rooster does not impede a hen's ability to lay eggs either unless he somehow damages or stresses her in his exuberance to breed.

So now you understand some of the normal behaviors of a hen as her hormones are produced and she gets ready to lay an egg. Of course, making that egg is quite a feat. But that hen sure makes it look easy.

TRIM A ROOSTER'S TOENAILS USING DOG OR CAT NAIL TRIMMERS AND THEN AN EMORY BOARD TO SMOOTH OUT ROUGH EDGES. FOR TRIMMING WITH MORE HORSE-POWER, USE A SMALL ROCK ATTACHMENT ON A ROTARY TOOL TO GRIND DOWN THE SPUR AND TOENAILS. BOTH EMORY BOARDS AND ROTARY GRINDERS TRIM BEAKS WELL. CHICKENS HAVE A KIND OF QUICK IN THE TOES AND SPURS SO HEED THEIR WARNINGS IF OR WHEN YOU REACH SENSITIVE TISSUE.

An Egg's Journey Through a Hen's Oviduct

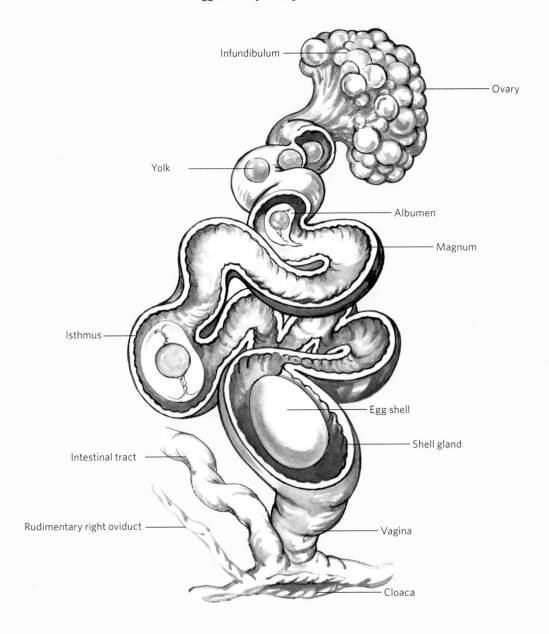

Infundibulum

Ovary

Yolk

Albumen

Magnum

Isthmus

Egg shell

Shell gland

Intestinal tract

Rudimentary right oviduct

Vagina

Cloaca

HOW A CHICKEN MAKES AN EGG

You reach into the nest box and wrap your fingers around a warm, just-laid egg. You pick it up, feel its warmth against your cheek, and then pop it into in your pocket. Back in the house, you heat up the fry pan and crack open your egg, staring at all of its curious parts and poking at the yolk with a spatula. This egg came from your happy hens; unhappy hens do not lay eggs so you must be doing something right. The yolk breaks open from the spatula's weight and spreads into the egg white. As it cooks, you wait patiently. When it is ready, you put it on a plate and serve yourself a fresh, egg-a-licious breakfast. The taste, well, those words are best left to your own imagination. But what about the journey that wonderful egg took to get to your breakfast?

The egg starts with the yolk. It begins small and is surrounded by a follicle necessary to transport the nutrients and pigments needed. Once ready—filled with sufficient nutrients and genetic material—the yolk descends into the oviduct, traveling first through a funnel-like section called the infundibulum. This is where it gets fertilized by sperm, if any has been stored. Next it ends up in the largest oviduct section called the magnum, where the yolk connects to the white, also called the albumen.

The next stop on the yolk's journey is the short, narrow isthmus, where the inner and outer shell membranes are added, the latter of which provides the foundation upon which the shell is built. The yolk and albumen spend seventy-five minutes in the isthmus. Once surrounded by the inner and outer shell membranes, the group continues on to the shell gland, where the shell gets added, hardens, and changes pigment color and where the albumen's four layers unfold. The egg spends more than twenty-four hours in the shell gland. Finally, the last layer, a thin protein mucus called the cuticle or bloom, seals the egg's pores to prevent invasion by bacteria. Now the egg is ready to be laid.

The egg is formed inside the hen with the small end facing downward. It flips around just before it is laid so that it leaves the hen large end first. The egg passes through the vagina and out through the cloaca or vent. If needed, a hen can hold onto an egg until a favorable place or situation to lay it. This is key for a wild hen avoiding a predator. In this rare circumstance, it is possible that a hen will lay two eggs within a twenty-four hour period. Whew! It takes all that to lay an egg. To all the hens in your coop: Take a bow. They have earned it!

Anatomy of a Chicken

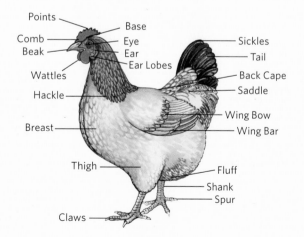

Points
Comb
Beak
Wattles
Hackle
Breast
Thigh
Claws

Base
Eye
Ear
Ear Lobes

Sickles
Tail
Back Cape
Saddle
Wing Bow
Wing Bar
Fluff
Shank
Spur

White Leghorn

SELECTING A BREED TO SUIT YOUR NEEDS

Which breed is right for you? Choosing your chicks may seem as simple as picking out the cutest one at the feed store, but some people want more information. The first question that always comes to minds is eggshell color. Eggs with different-colored shells do not have any nutritional difference from white-shelled eggs; the shell is the only place pigment is added. The contents are exactly the same.

The American Poultry Association puts out what's called the *American Standard of Perfection*. This publication provides color, type, and size guidelines for all accepted chicken breeds and is used as a reference by many chicken owners.

Select your chicken breed based on the environment you can provide and on their behavioral tendencies. For example, if you have harsh winters and a small coop, choose a cold-hardy breed with a mellow disposition. A dual-purpose breed, one used for both meat and eggs, would be an excellent choice. We suggest breeds from the American and English classes of the American Poultry Association if you need calm, cold-hardy birds.

Bantam breeds are about one-fifth the size of standard chickens. The benefits are that you can fit more of these chickens in a smaller space, and they eat much less food than their full-size counterparts. Bantam eggs are about one-third the size of standard eggs. There is one notable concern: They fly better and more readily than most larger chickens. Their wings need to be clipped or their runs need to be covered to prevent escapes.

If you instead choose a laying hen known for its high production levels such as a White Leghorn, understand that it is an extremely active breed with high levels of aggression. Alternatively, a breed selected for its rapid growth and deemed a broiler chicken (one raised for its meat) typically isn't aggressive. Breeds developed in warmer regions of the world such as the Egyptian Fayoumi may not perform well in cold environments unless you modify the coop to meet their needs through added insulation, for example.

Although many people tend to choose chickens based on pretty feathers or eggshell color, keep in mind their behavioral needs as you make your decision. Also, make sure you take into account your city, state, and federal laws, which we discuss in chapter 3.

Silver Sebright bantam rooster

Look to the Earlobe

Which hens lay brown-shelled eggs? This is always a first question because brown-shelled eggs are perceived as fresher, more natural, and if nothing else, at least different from grocery store eggs. The rule of thumb is simple (but not perfect) and has to do with a hen's earlobe, the fleshy spot just beneath the feather-covered ear. A red earlobe means brown-shelled eggs and white earlobes mean white-shelled eggs. There are chickens that lay blue-green shelled eggs and dark brown-shelled eggs, but we'll talk about them later.

Javas

Breed Terminology

Large fowl chickens are organized by region of the world. A "class" is a group of breeds organized and named by the region of origin in the world. Examples are American, English, Asiatic, Mediterranean, and so on. A "breed" is a subdivision of a class. Some breeds in the American Class include Plymouth Rocks, Dominiques, Javas, New Hampshires, and so on. A subdivision of an individual breed is called a "variety." Often a variety is based on plumage color, but comb type or presence of a beard and muffs can also determine this. There are many varieties within a breed and many breeds within a class.

Breeds with Brown-Shelled Eggs

To get you started on your backyard chicken journey, we present an introduction to each breed, along with a picture to help you decide which will enter your flock. Many breeds offer multiple varieties, so you still need to do more research to make a final decision. We suggest many of the breeds here for both their ability to lay brown eggs but also their hardiness against the cold. This list discusses large fowl only, as the color of eggshells and cold-hardiness of bantams for each breed can differ significantly from their large fowl counterparts.

PLYMOUTH ROCK

When people imagine an at-home flock, they typically think about the now-iconic Plymouth Rock. These are excellent dual-purpose birds in that the males are raised for meat and the hens are kept for eggs. Many people also recall hearing about Barred Rock chickens. This is a mispronunciation or abbreviation of the whole name, which is Barred Plymouth Rock.

Appearance

The hens weigh about 7½ pounds (3.4 kg) and have yellow skin and yellow legs and toes. The breed, developed in the United States in the nineteenth century, is believed to be the result of a cross between a Dominique male and Java or Cochin female. The first variety was a Barred bird, with White sports (what some varieties are called) occasionally making an appearance. White Plymouth Rocks were the next variety accepted into the *American Standard of Perfection* reference. Female White Plymouth Rocks were used to create

Barred Plymouth Rock

Rhode Island Red hen

the commercial strain of meat birds raised as Cornish crosses or broiler chickens. They were key in developing broilers because they were not only good-size birds but also better egg layers than the Cornish. Other varieties of the breed still exist: Buff, Silver Penciled, Partridge, Columbian, and Blue.

Personality

The personality of this breed is best described as calm. These birds are good with children and people new to chicken-keeping. They are also sweet and genuinely want to be around their caretakers and other pets or chickens. They are curious and outgoing, the type of hens that will check out every new item you bring to the coop or backyard. They are independent, too, able to be left unsupervised and to fend for themselves quite well if left to roam in the barnyard.

RHODE ISLAND RED

Developed in New England, in the region between Narragansett Bay, Rhode Island, and Buzzards Bay, Massachusetts, the Single Comb variety of this breed was recognized by the American Poultry Association in 1904. In 1905, the same happened with the Rose Comb variety of the Rhode Island Red. Both developed from a cross between a Red Malay Game, Leghorn, and Asiatic stock.

Appearance

Rhode Island Reds are a lustrous, deep red, almost mahogany. Even the fluff of the bird should be a rich, intense red. The tail should be black. The skin of the bird is typically yellow and the legs and toes are a rich yellow tinged with reddish horn color. The beak also is a reddish horn color. The hens weigh about 6½ pounds (2.9 kg); pullets are about 1 pound (0.5 kg) lighter. This breed was originally used as a market bird and as such, has retained its capacity as an excellent dual-purpose breed, perhaps the best kind for egg production.

Personality

The personality of Rhode Island Reds and Rhode Island Whites (the next breed listed) are so similar that we'll discuss both in this paragraph. They are cold-hardy hens that are as sweet as they come, curious and active birds that will readily investigate new items or till a garden at your request. They are outgoing and can get along well with well-trained family pets. They are an excellent choice for the beginning flock owner and are very good with children. As with any new chicken, taming—through regular handling from the time the birds are chicks—will need to take place to ensure a good experience for both child and chicken.

Rose Comb Rhode Island White Cockerel

Rhode Island Red chick

RHODE ISLAND WHITE

Many people do not realize that the Rhode Island White breed even exists. It came after its more famous counterpart, the Rhode Island Red, and it has a slightly different lineage. It was developed in Rhode Island in the late nineteenth and early twentieth centuries, the result of a cross between Partridge Cochins, White Wyandottes, and Rose Comb White Leghorns. Like Plymouth Rocks, these are dual-purpose birds, with the males used for meat and the hens for eggs. Rhode Island Whites, which are rare, come in only one variety: Rose Comb.

Appearance

The hens are about 6½ pounds (2.9 kg), and have yellow skin and yellow legs and toes. Will this breed be the jewel in your coop?

Personality

For a description of the personality, see Rhode Island Red.

Buckeye

BUCKEYE

The Buckeye breed has seen a resurgence in popularity over the years. It is critically endangered, but through the efforts of show poultrymen and -women, backyard flock enthusiasts, and breed preservationists, the breed is making a comeback.

Appearance

The Buckeye has the distinction of being the only breed developed by a woman, Nettie Metcalf. Named for her home state (Ohio), the Buckeye has plumage that resembles the color of the buckeye nut, similar to that of Rhode Island Reds. Its legs and toes are yellow. A cross between the Dark Cornish, Black-Breasted Red Game, Buff Cochin, and Barred Plymouth Rock, this breed is stout and wide, but not to the same degree as the Cornish. This breed comes in only one variety and has a pea comb. Hens weigh approximately 6½ pounds (2.9 kg).

Personality

The personality of the Buckeye is not unlike the three previous breeds mentioned. These birds are sweet and depending on the strain purchased, a bit independent or standoffish. Also good with children, they are agreeable birds for poultry showmanship (the unique history of the bird offers an excellent talking point).

New Hampshire rooster

Personality

New Hampshires can be aggressive, competitive, and dominant within the flock. They are curious birds and quite independent. They also are likely to wander off and lay eggs in a private clutch rather than the nest boxes you kindly provided (although this can be said about other breeds as well).

CHANTECLER

The Chantecler, the only breed from Canada, was bred specifically to handle the region's cold winter conditions. The goal with these animals was to develop birds with a small comb and wattles that laid well through the winter months and had good health and vigor. Bred by monks, lead by Br. Wilfred Chatelain from 1908 to 1918 at the Oka Agricultural Institute in Quebec, the breed came from two separate crosses.

Specific lineages for hens and roosters were created, then mated together, with subsequent selections done to create the bird that we know today.

Appearance

The Chantecler is one of the only chickens with a cushion comb. The hens are 6½ pounds (2.9 kg) and the breed comes in two varieties: White and Partridge. Yellow skin, legs, toes, and beak round out the White variety. The Partridge variety has a dark horn beak that shades to yellow at the point.

Personality

A Chantecler is a classic, cordial, but independent farmyard chicken. The hens lay a fair number of eggs and are often a favorite of small flock owners due to their sweet nature.

NEW HAMPSHIRE

Nowadays, we think of New Hampshire as an egg-laying breed, but its history and selection lean it more toward a meat bird. Therefore, it is truly dual-purpose breed. The lineage stems entirely from the Rhode Island Red. It moved from the Rhode Island area up into the New Hampshire region, where it was selected for rapid growth, fast feathering (handy in the cold northern parts), early maturity, and vigor. Eventually an entirely new breed developed from the farm stock maintained by New Hampshire's farmers.

Appearance

These birds are brown egg layers, and they also produce a good-sized carcass. The color of the pinfeathers do not detract, as they are a reddish buff color. They are fair, but not the highest producing, egg layers.

Dominique hen

DOMINIQUE

This breed has the distinction of being the United States' oldest. As such its lineage is a bit hazy. These birds are dual-purpose and come in only one variety, Barred.

Appearance

To the untrained eye, this breed looks similar to a Barred Plymouth Rock; the key is in the comb. The Dominique has a rose comb while the Barred Plymouth Rock has a single comb. The Dominique also carries its tail at or higher than a forty-five degree angle above horizontal while the Plymouth Rock carries its tail at thirty degrees. The beak, toes, and legs of the Dominique are all yellow. The breed in general is a bit smaller, with hens weighing in at 5 pounds (2.3 kg). The hens are fair layers of brown-shelled eggs.

Personality

Dominique hens are sweet and curious. Like Plymouth Rocks, they like to spend time with flock mates and human caretakers alike. They will check out every new item you bring to the coop or the backyard. They are sweet with children and are often the bird of choice for the youngest flock owners in your family because the hens are lighter. These easy-to-pick-up hens do not fuss much and get along well with other pets and livestock. But there is a note of caution: Given their slightly smaller size and docile nature, these birds can be pushed around a little at feeding time by other, larger flock members.

Silver-Laced Wyandotte

WYANDOTTE

If ever a bird or breed got *ooohhhs* or *aaaahhhs* from novice poultry enthusiasts, it is the Wyandotte. This beautiful breed was developed in New York State. The original variety, Silver Laced, is considered the parent of the Wyandotte family. It is sometimes called an American Sebright or Sebright Cochin. Although the exact breeds used to create the Wyandotte remain unknown, it is suspected these include the Dark Brahma and Silver Spangled Hamburg. This is also evidenced by the fact that Wyandottes have rose combs.

The Golden Laced variety came next among the Wyandottes, from a mating between a Silver Laced Wyandotte female and a crossbred Partridge Cochin/Brown Leghorn cockerel. Both the White and Black varieties came from sports of the Silver Laced variety. The Buff variety descended from Rhode Island Reds of poor type and color. Thus these birds were crossed with other breeds and the Wyandotte to create the Buff

variety. The Partridge variety consisted of two independent efforts in its creation. The last Wyandotte variety is the Columbian. Often overlooked, yet an elegant and stunning variety, the Columbian was named after the Columbian Exposition of the World's Fair held in Chicago, where it made its exhibition debut.

Appearance

This breed is often recognized as the jewel of the chicken coop, though it is quite often overlooked due to the popularity of other more common and well-known breeds. It is the discerning flock owner who chooses the Wyandotte. This fairly good egg layer is an average-size bird weighing in at about 6½ pounds (2.9 kg). The rose combs are small and are not prone to freezing. Although not to be kept as breeders, an occasional single comb sport can be seen in offspring.

Personality

The hens are sweet, curious, good mothers, and well behaved with children. They walk with you and cluck or talk with great frequency. Thus, many owners view them as understanding a conversation. The beauty and good disposition of this variety usually garners the most interest and comments from anyone visiting your flock.

DELAWARE

Don't confuse the Delaware breed with the Delaware Blue hen. The two breeds, although both exceedingly rare, differ vastly in personality and function. Delaware Blues descends from fighting stock, are very athletic, and are not recommended for a beginning flock owner. Delawares, on the other hand, were the first attempt at creating a broiler or meat chicken.

Appearance

The cross between Barred Plymouth Rock males and New Hampshire females yielded oddly colored sports. The most striking feature of the Delaware is the barred hackle and tail feathers, while the rest of the bird is pure white. The hens weigh approximately 6½ pounds (2.9 kg) and are good layers. The males make fine broiler chickens. Another unique feature is that backcrossing of the Delaware to other breeds yields sex-linked chicks. For example, Delaware males bred to Rhode Island Red or New Hampshire females yield chicks with the markings and plumage of the Delaware. But Delaware hens bred to male Rhode Island Reds or New Hampshires yield the sex-linked chicks. Male chicks will have the Delaware pattern and female chicks will have the solid red feather pattern and therefore can be identified at hatch.

Personality

The Delaware yields even-tempered hens that are independent birds designed for foraging in the barnyard. These hens are easily tamed and therefore become good companions for the gardener. They have bright, inquisitive eyes and are often a quiet breed. Not always willing to share what's on their mind, they instead tend to quietly follow caretakers around the yard while tilling and turning the soil. A Delaware will still alert you to problems such as predators. They are very good mothers and good with children.

Buff Orpington

ORPINGTON

Orpington is a large breed that hails from the town of Orpington in County Kent, England, so it falls into the English class. It is derived from a cross between a Black Langshan, Black Minorca, and Black Plymouth Rock. The popularity of these birds in the United States was solidified when they were shown in great numbers at the 1895 Madison Square Garden Show in New York. Their size makes them excellent as dual-purpose birds as both the hens and roosters can be used for meat.

Appearance

These birds carry many feathers held loosely from the body, making them appear even larger than they already are (hens weigh 8 pounds [3.6 kg]). These soft feathers are often solid in color as the hens come in White, Black, Blue, and Buff. (The Buff variety is the most popular, although White comes in a close second.)

Personality

Orpingtons are sweethearts. They walk with you and talk with you about absolutely everything. They have reddish bay eyes that beckon for conversation and watch your every move. They are docile and excellent with children, although quite sizable and difficult for children to carry. They are not aggressive and are excellent mothers. They do, however, tend to go broody (see chapter 5 for more on hens going broody) and are good incubators. Despite their size and given their sweet nature, these birds can be the lowest in the peck order, so be sure to select even-tempered breeds to raise with them. Or raise a flock entirely composed of Orpingtons.

Australorp hen

Buff Orpington hen

AUSTRALORP

The Australorp is a breed of chicken that lays large numbers of brown-shelled eggs. It is also the only breed from Australia and is in the English class, dating back to the days of the British colonies. The breed was developed out of Black Orpington stock and was known as a production-bred Australian Black Orpington.

Appearance

The only breed they come in is Black, but they should have a lustrous green sheen to their feathers. The hens weigh 6½ pounds (2.9 kg). They are great egg layers. In fact, in one Australian test, a hen was recorded to have laid 364 eggs in 365 days!

Personality

The breed is sweet, curious, social, and a good forager. These birds do not tend to go broody and make an excellent selection for the farmer wishing to start a small enterprise with brown-shelled egg layers. They also have dark-colored eyes that appear quite inquisitive.

THE CHICKEN WHISPERER'S GUIDE TO KEEPING CHICKENS

SUSSEX

The Sussex breed hails from Sussex County in England and is cold-hardy. It is an old breed, typically kept as dual-purpose. Sussex County was known for producing high-quality table fowl. With the production of a fowl for meat also came the demand for that bird, hence the breed also has good egg-laying characteristics.

Appearance

The feature of this breed that interests most flock owners is its plumage. The Speckled Sussex has mostly reddish feathers, each tipped at the end with a white spangle. Between the reddish bay on each feather and the spangle lies a black bar. This interesting feather pattern has increased the Speckled Sussex's popularity among small flock owners. Do not forget about the other two varieties in which this breed comes. Red Sussex are a deep mahogany color, with pinkish-white legs and toes. The Light Sussex has a color pattern that mimics the Columbian Plymouth Rock.

Personality

Soft-spoken, yet independent and inquisitive, the Sussex is an excellent choice for small flocks. These birds are good with children. Plus, they are active and will make a beautiful addition to your backyard flock. They are good setters and willingly sit on eggs and brood chicks.

Sussex

High Energy, White-Shelled Egg Layers

Many white-shelled egg layers also tend to be breeds that hail from the world's warmer regions. Some of the breeds mentioned in this chapter are also high-energy birds. This means they aren't necessarily as friendly and may have a tendency toward flightiness. Of course, this does not apply to every bird in every case.

Ancona

LEGHORN

The Leghorn is known for its high egg production. In fact, it is considered the best egg-laying breed and the stick by which all others are measured. Though the Single Comb White Leghorn, (see page 24), is the foundation breed for the modern commercial egg-laying industry, Leghorns do come in many beautiful varieties. The breed originates from the Livorno region of Italy and is therefore part of the Mediterranean class. These are hardy birds, but not necessarily cold-hardy. They eat one-third less than their brown-egg laying counterparts, yet lay more eggs, making them a logical economic choice for small businesses in need of white-shelled eggs.

Appearance

Leghorn hens have large combs that flop over one side of their heads, either partially or fully obscuring one eye. There are at least ten Single Comb varieties, including Light Brown, Dark Brown, Buff, Red, Columbian, and Golden Duckwing, to name a few. There are fewer Rose Comb varieties, but they include Light Brown, Dark Brown, White, Buff, and Silver. These birds are small-bodied with large combs, making winter care more of an effort for their caretakers. Leghorns, as a breed, are known for distinct lack of broodiness.

Personality

Leghorns are flighty and noisy. Many first-time flock owners who start with Leghorns coupled with brown-shelled egg layers prefer the calmer tendencies of the latter birds. Easily frightened, the high-strung Leghorns are watchful and prefer to keep their distance from new items and people. They are strong fliers given their light body frame and small weight of 4½ pounds (2 kg). They are aggressive if overcrowded and tend to become the dominant members of a mixed flock. Given the right conditions, Leghorns are good layers but are not a preferred first choice for small or backyard flock owners. Reserve the Leghorn for your second or third flock—when you may want more of a challenge.

ANCONA

For several decades, the Ancona was considered just another variety of Leghorn. Just like the Leghorn, this breed also comes in Single and Rose Comb varieties. Descended from hens of central Italy, the breed was named for the port from which it was sent to Great Britain. It is in the Mediterranean class of chickens and is not cold-hardy.

Appearance

These hens weigh 4½ pounds (2 kg) and are known for their lack of broodiness. Though there is only one Ancona color variety—Mottled—the breed is gaining popularity because of its unique look. Each feather is black with white tips. Like Leghorns, these birds also have a large comb that flops to one side of the head. At one

THE CHICKEN WHISPERER'S GUIDE TO KEEPING CHICKENS

time the breed was known for its excellent egg production, but now these birds are strictly ornamental.

Personality

The Ancona, quite a flighty breed, can get as noisy as the Leghorn. The hens are excellent egg layers but are not good choices for backyard flock owners. They are not good with children although they are great yard or garden foragers. These birds are strong fliers, so clip one wing on each bird to prevent roosting in trees.

ANDALUSIAN

As its name implies, this breed hails from Spain's Andalusian region. As such it belongs in the Mediterranean class and comes in a most beautiful variety: Blue.

Appearance

Blue Andalusians are blue slate or slate gray. The Blue Andalusian is a perfect example of color genetics in poultry. The original bird—as is occasionally done with today's birds—was bred from a cross between a Black bird and its White sport offspring. A result of incomplete dominance, the blue color is unusual and difficult to maintain. These birds are slightly larger than Leghorns, weighing in at 5½ pounds (2.5 kg).

Personality

These light-bodied, high-energy birds are excellent foragers. They are elegant and inquisitive yet initially standoffish. Once they get to know their handlers, they like spending time with them in the garden and are extremely alert to changes, new people, or predators. They also can be aggressive toward other hens and tend not to go broody. A decent choice for small flock owners who want white-shelled eggs, these birds are intelligent and relatively easy to tame.

Blue Andalusian hen

WHEN TWO BLUE ANDALUSIANS ARE MATED, ONE-QUARTER OF THE OFFSPRING WILL BE BLACK, ONE-QUARTER WILL BE WHITE, AND THE REMAINING HALF WILL BE BLUE. THE WHITE AND BLACK OFFSPRING, IF MATED TOGETHER PRODUCE MAINLY BLUE BIRDS. BUT THERE ARE COMPLEX COLOR GENETICS AT PLAY. THEREFORE NOT TOO MANY BREEDERS CAN MAINTAIN HIGH-QUALITY BLUE-COLORED BIRDS. THOSE WHO HAVE BEEN SUCCESSFUL WITH THIS BREED RAISE BIRDS WITH BEAUTIFUL BLUE LACING AROUND LIGHT BLUE-GRAY FEATHERS.

Hamburg

Dorking

HAMBURG

Many think the Hamburg breed comes from Germany, as the name would imply, but actually, the Hamburg originated in Holland. Breeders in Germany and Great Britain, however, refined the breed to what it is today: Pheasantlike in body shape and behavior. The breed is known as a good layer—giving it the unique nickname of "The Dutch Everyday Layer"— but laying medium size eggs.

Appearance
Hamburgs have beautiful plumage and come in six varieties: Golden Spangled, Golden Penciled, Silver Spangled, Silver Penciled, White, and Black.

Personality
This is a high-strung breed with a rather wild and flighty temperament. It is not recommended for first-time flock owners. These birds are noisy and excellent fliers, so prevent tree roosting and escape by clipping one wing. They are excellent at foraging on their own and are independent, preferring to spend almost no time with humans.

DORKING

Originating in Italy, this ancient breed came with the Romans to Great Britain, where most of its development took place. This breed is in the English class and is somewhat cold-hardy. These birds do have larger combs and therefore require a bit more protection from frostbite, but this dual-purpose breed is better known for its other unique features.

Appearance
The body is squat and square, with Silver-Gray hens having a beautifully varied plumage. (This breed also comes in the White variety.) These birds have five toes and the hens weigh about 7 pounds (3 kg). It's one of the few breeds whose birds have red earlobes but actually lay white-shelled eggs.

Personality

The breed is docile in temperamentand the hens are good foragers and tend to go broody, making them good for sitting on eggs. Keep this detail in mind if you intend to keep the breed for egg laying. Hens that go broody do not lay eggs. This is a sweet breed that spends plenty of time with you in the yard but yet is independent enough that the birds can be left to work on their own in the garden.

Other Eggshell Colors and Accoutrements

Although we've mentioned many chicken breeds thus far, there are still many more out there with unique or intriguing features. For example, many breeds become famous for unusual feather accoutrements or eggshell colors. If you wish to sell your extra dozens, pay special attention to the breeds in this section. They will certainly put your flock on the map!

AMERAUCANA

This breed, a derivative of the Araucana, also produces blue-green or turquoise shelled eggs. The Ameraucana came about to create a dual-purpose breed that not only had the unique eggshell color but also good meat production qualities. The Ameraucana is more easily obtainable than the Araucana and therefore costs less. Note the correct name spelling here. Many hatcheries and feed stores tend to misspell the name as "Americana".

Appearance

Ameraucana does not have ear tufts. Instead, the breed has a beard and muffs under the beak in place of wattles. Through crossbreeding the Araucana with other breeds, a tail also emerged in this breed.

Ameraucana

Personality

Many flock owners find this sweet, curious breed quite enjoyable in the yard. Much like the Plymouth Rock, these hens show you how much fun your own backyard can be! They are excellent with children but grow larger than their progenitors, weighing in at 5½ pounds (2.5 kg). They are good layers and can make good mothers.

Araucana

Cochin

ARAUCANA

This is the breed that lays the famous turquoise eggshell. Unfortunately, more often than not, the Araucana gets confused with the Ameraucana. The distinction *must* be understood as the two are very different. One gave rise to the other! The Araucana is the only breed to hail from South America, specifically Araucana, Chile. Little is known about its history, but it is a good dual-purpose breed. The blue-green eggshell trait is dominant and will persist in the next generation even if crossed with another breed.

Appearance

The original breed should be rumpless, or without a tail. This happens because these birds are missing the last segment of their spine where tail feathers typically sit. Another unique feature of the Araucana is that it has an extra bone in each ear, upon which grow feather tufts. An Araucana should have wattles beneath its beak, too.

If you purchase what you thought were Araucanas and the birds grow tails, you likely got Ameraucanas instead. Buyer beware: You did not get pure stock! True Araucanas come in Black, Black Red, Golden Duckwing, Silver Duckwing, and White varieties.

Personality

Araucana hens are sweet, curious, and active foragers. They are definitely worth the investment, but insist on quality stock. They are excellent birds for children as they are light hens (only 4 pounds [1.8 kg]) and provide excellent talking points for showmanship.

COCHIN

The Cochin is a very old chicken breed that came from China. It falls into the Asiatic class. The breed arrived in Great Britain and America in 1845, where the birds caused a Cochin craze: Everybody wanted them, everyone bred them. Today's Cochins are mainly used for ornamental purposes and will certainly be a keen talking point if added to your flock. They are not fast-moving hens, but they will forage with other chickens in the yard.

Appearance

These birds have feathered feet and shanks, making them prone to muddy, dirty feathers after it rains; keep them in confinement on wet days. They are loose-feathered birds that look like large, round puffballs with so many feathers that no light can pass between the legs. These birds come in Buff, White, Black, and Partridge, with high-quality Partridge being the most difficult to find and Buff remaining the most popular. The Cochin is cold-hardy although it has a large comb, meaning take extra care on cold days. These large hens weigh about 8½ pounds (3.9 kg).

Personality

Cochins are known for their extraordinary mothering characteristics. They are prone to broodiness and though they are not profuse egg layers, they again and again raise broods of chicks to adulthood. Be aware though that you'll need to force some of these hens off the nest to encourage eating and drinking. Also, they are sweet and curious birds. Despite their large size, they are heartily recommended for children. The hens are patient, kind, and talk to their caretakers; like the Orpington, this breed is a thoughtful clucker and all you need do is respond—you may even see a little smile form on their beaks!

FAVEROLLES

The Faverolles breed comes from the village by the same name in France, and it is in the Continental class. It is a result of many years of breeding between the Houdan, Dorking, and several Asiatic breeds. These birds were bred for winter egg production and are cold-hardy.

Faverolles

Appearance

Because the breed has fairly loose feathering, these birds appear larger than they actually are. The hens weigh about 6½ pounds (2.9 kg) and have a beard and muffs beneath the beak. Members of this breed have a fifth toe along with feathered shanks and a small single comb. They lay lightly tinted brown- to tan-shelled eggs. Chiefly used for ornamental purposes, this breed has beautiful plumage. The Salmon Faverolles, for example, has beautiful pastel plumage that makes it a popular selection. The breed also comes in White.

Personality

Though quiet and sweet birds when it comes to being around children, these hens are gregarious and curious about their surroundings. They also are excellent showmanship birds. They always appear alert as their tails rest at fifty degrees above horizontal. This is a good breed for a beginning flock owner, but do not expect high levels of egg production.

Brahma rooster

Light Brahma hen

BRAHMA

Brahma, developed in China, was originally called Chittagong, Gray Shanghai, and Brahma Pootra. The name was later shortened to the Brahma. It is considered a cross between the Malay and Cochin in India.

Appearance

The first varieties of Brahma were Light and Dark, followed fifty years later by the Buff. These birds should not have feathering as loose as the Cochin but should have feathered shanks and feet. They also need especially deep bedding to preserve the feathers on their feet (not unlike the Cochin, which should be kept out of the mud as well). The hens weigh 9½ pounds (4.3 kg), have a pea comb, and are decent egg layers. Their deep-set eyes makes them appear stern, but do not let that small detail deter you from adding this stately bird to your flock. They make good mothers but are mainly kept for ornamental purposes.

Personality

They are well known for their size and height, yet gentle in nature. Even Brahma males are docile enough for children to handle (though the roosters can weigh about 12 pounds [5.4 kg], so it is not recommended). Patient with children, Brahma make excellent showmanship birds. They are not the fastest moving flock members and prefer to mosey around the garden instead. They are quiet except when purposely directing clucking comments at their caretakers. These soft-spoken birds are sure to be the biggest in your coop.

BARNEVELDER

The Barnevelder is one of the most recent breeds accepted into the *American Standard of Perfection*. It hails from Barneveld, Holland, and is in the Continental class. It is famous for its dark-brown eggshells.

Appearance

The only recognized variety of this breed is the Double-Laced Partridge. Each feather should be reddish brown with a black lacing that has a lustrous, green sheen. The hens weigh 6 pounds (2.7 kg) and should have single combs. There are not many breeders of the Barnevelder and it remains a hard-to-find specimen.

Personality

The hens are independent yet curious and sweet. They are good yard and garden foragers. Barnevelders should be encouraged to lay in nest boxes because their dark-shelled eggs can be difficult to see or find in hidden nests. They are good hens for those starting flocks and produce highly marketable eggs.

WELSUMMER

The Welsummer, another recent breed, comes from Welsum, Holland and is also in the Continental class. It is a composite of several breedings between the Partridge Cochin, Partridge Wyandotte, and Partridge Leghorn. The Barnevelder and the Rhode Island Red were added to the mix to create the breed we know today. The Welsummer's claim to fame comes in the form of its dark-brown eggshell color.

Welsummer

Appearance

The hens weigh approximately 6 pounds (2.7 kg) and have medium-sized single combs. The striking plumage of this breed puts it in the same category as the Silver-Gray Dorking, with a series of colors all over the hen's body: with a brown and yellow head, a pinkish salmon breast, and the remainder of the bird a beautiful shade of speckled brown.

Personality

The Welsummer lacks broodiness. However, it is a good forager and extremely independent. The breed is a good egg layer, although not likely the most proficient one in your flock. The Welsummer, though a tad skittish, can be tamed so that it is more approachable. These birds are active, ready to bound around the yard in search of worms and grubs in the garden or grass. With time, Welsummers accept new people into their circle, producing strong bonds. Worth a little more work, a Welsummer makes a welcome addition to your flock.

Marans hen

MARANS

The Marans is the newest breed on the block and therefore, there's not much information published about its history or body conformation.

Appearance

They are known for extremely dark-shelled or chocolate-colored eggs. This tends to be an excellent selling point for those eggs. Customers buying mixed or whole dozens often are willing to pay a bit more for the Marans' dark-shelled eggs. Feathered shanks appear to be an important characteristic, but further details on conformation will need to wait.

Personality

These are docile, mellow birds with sweet personalities.

FINDERS KEEPERS

Remember, all flock members are unique individuals and may not conform to the personalities or descriptions here. Now that you know how the egg forms, perhaps you have more appreciation for the hard work that goes into the process of laying an egg. Go ahead; make deviled eggs, crème brulée, or perhaps a little eggnog. No matter what you prepare, thank your hens every so often with a little extra treat.

This breed guide should help you decide which breeds meet your needs. Perhaps you are interested in a breed not mentioned here. That's fine. Make sure you find a breeder who can teach you more about that interesting breed. And for breed preservation—if that is part of your goal—always source your birds from reputable breeders. Also, there are many more breeds from which to choose if you delve into the world of bantam (miniature) chickens.

Giving a new hen a safe home in your coop will ensure that you continue to get eggs from happy hens.

Marans rooster

CHAPTER

3

BACKYARD OUTLAWS:
ARE CHiCKENS ALLOWED WHERE YOU LiVE?
WHAT SHOULD YOU DO iF THEY AREN'T?

BEFORE STARTING your backyard flock, make sure chickens are allowed where you live. Start hyperlocally by learning about regulations such as those from your homeowners association, then move outward. Check with city, state, and federal laws, too. Many major cities, for example, seem to allow backyard chickens, but the smaller ones do not. (The only conclusion we can draw as to why relates to the notion that larger cities are typically older, meaning at their start, residents likely kept chickens and other livestock for food supplies.) Solid resources where you can find out about chicken-related laws near you include www.municode.com and www.thecitychicken.com. Also, learn the terminology that applies to your area. Some laws state that only domesticated animals are allowed. Technically, chickens were domesticated at one point in time, some 6,000 to 10,000 years ago. Other laws may state that only household pets are allowed. One definition of pet is an animal kept for companionship and pleasure. Chickens could fall into that category. Finally, some laws classify poultry as livestock and apply to it all accompanying livestock regulations.

[Chickens will always find a new social gathering place.]

GETTING LAWS CHANGED

If you find out that keeping backyard chickens is not allowed where you live, you have three choices. Don't keep backyard chickens, keep them under the radar, or work toward trying to get the laws changed so you can keep them legally.

Before heading off to that city council meeting to get the backyard chickens laws changed, you need to arm yourself with information. People tend to gripe about problems they think accompany backyard chickens such as noise, smell, rodents, disease, or decreased property values. Here's what you need to know:

The Truth About Noise

The stereotype that a rooster crows early every morning (as well as throughout the day) is true. Therefore, complaining about a rooster's noise is valid, especially if the bird likes to sound his alarm well before sunrise.

If you are a late sleeper, roosters are not for you. Yes, it's true. They *do* cock-a-doodle do!

But here's your argument: Most urban chicken keepers want backyard chickens for the endless supply of fresh eggs, not to breed the birds for chicks. Guess what? You don't need a rooster to get eggs from hens. In fact, hens tend to lay better without a rooster around to disturb their routine. And in contrast to their male counterparts, hens are quiet, sleeping soundlessly through the night.

If you do want a rooster, try a different tack: Obtain noise-complaint statistics from the local law enforcement of a town that does allow roosters. Chances are that complaints about barking dogs and loud music far outnumber those resulting from crowing roosters.

The Truth About Smell

Smell is another complaint often brought up when discussing chickens. Yes, chickens can produce an odor—just like dogs, cats, rabbits, hamsters, gerbils, even people—if not cared for properly. The smell most people think of comes from commercial chicken houses holding tens of thousands of chickens. That's what most people think of when you say you want chickens in your backyard. The reality is that you'll probably keep anywhere from six to twelve laying hens in your backyard, no more. Plus, the best way prevent your chicken coop from smelling is to keep it clean.

Laws to Consider

In addition to whether chickens are allowed in your town, there are other broad requirements you should learn before investing in backyard birds. Here are a few to think about:

- → Are there height requirements for the coop?
- → Are there distance requirements for the coop (e.g., how far it must be from the nearest neighbor) or property line?
- → Is a coop considered an accessory structure? Whether it has this title could mean different regulations.

The Truth About Pests

Keeping chickens brings mice and rats, right? No. Chickens themselves don't attract mice and rats, but rather the rodents come for a food and water source. In that vein, a backyard chicken feeder is no different than a typical wild birdfeeder when it comes to providing food for mice and rats. Same goes for a chicken waterer; as a water source for mice and rats, it's no different than any outside potted plant. In all likelihood, you have mice and rats outside your home right now that you just don't know about—and so do your neighbors.

The Truth About Disease

People tend to express concern about diseases backyard chickens might bring to the area. In 2007, a worldwide outbreak of avian influenza lead to a flurry of questions about the risks of this disease and keeping backyard chickens. In the United States, the Centers for Disease Control and Prevention (CDC) addressed this issue. When asked whether it was safe to keep a small flock of chickens, the CDC responded that at that time, in the United States, there was no need to remove a flock

of chickens because of avian influenza concerns. The agency also said that the U. S. Department of Agriculture keeps a close watch on poultry and poultry products for avian influenza viruses and other infectious disease agents.

Few diseases that affect chickens also cause illness in humans, and the diseases that do often result from unclean living conditions for the flock. The take-home message is simple: Always keep your flock's living conditions clean and free from mud, built-up feces, wild birds, and rodents.

A veterinarian inoculates a rooster against avian flu virus in 2007.

The Truth About Property Values

ANDY'S ANECDOTE

Time and again, I've heard about people who want to keep backyard chickens where they're not allowed. They say that because they have a strong relationship with their neighbors, they'll have no difficulty getting permission to start a flock. This may not be the best plan. Why? Sometimes your neighbors are the very people who report you to the authorities. They are the citizens who live in closest proximity to your flock and are the ones who may have a change of heart.

Don't just assume because you can keep chickens that you can keep other species of farm animals—there are different regulations and criteria for different breeds, for their safety as well as yours.

Some who oppose backyard chickens do so because they believe the chickens will decrease the value of their home and others in the neighborhood. This is a hard argument all around; unless a family has specific evidence—a signed letter from the homebuyer, for example—showing that the chickens next door caused the house to sell for less money, it's nearly impossible to prove.

The Truth About Other Farm Animals

Allowing backyard chickens does not mean the town also has to allow goats, pigs, or cows. I have never heard of a town that allowed backyard chickens and now has a goat, pig, or cow problem.

ENFORCEMENT CONCERNS

Let's say you convince your neighbors and the rest of the town that keeping chickens won't be noisy or smelly, and it won't bring in disease or drop property values. You still need to convince city officials that there are minimal costs associated with enforcing backyard chicken regulations.

Look closely at the laws and ordinances already in place. There are likely more than enough on the books right now to enforce any problems an irresponsible backyard chicken keeper might pose. What if a rooster is crowing at 4:00 a.m.? What if a dog was barking or a neighbor was playing loud music at the same hour? The aggrieved can file a noise complaint. What if a chicken gets loose in the neighborhood? Well, what if a dog gets loose in the neighborhood? Go after it and bring it home. Or for instance, what if the chicken run starts to smell? People can address this complaint as they would if a dog pen or compost pile started to produce a stench: by talking to the person or home generating the odor or by complaining to the local government.

Talking-Point Tips

When getting in front of a room to present your backyard chicken case, try these tactics:

Speak first. If your planned remarks cover and debunk myths and stereotypes about keeping backyard poultry, others who had planned to speak may no longer have solid arguments.

Thank the room. Start with something such as, "Thank you for the opportunity to speak today about the growing urban chicken movement." By killing the room with kindness rather than beginning defensively, you may knock detractors off their track.

Present your case effectively, with facts. If you do this, you force those against backyard chickens to either ignore facts or vote in your favor.

Address complaints and questions one component at a time. For example, say someone stands up to explain that he's had experience raising chickens, that he had to spend all of his free time cleaning out the coop, and that to protect the animals, he had to stand guard with a shotgun. Instead of getting defensive, ask him how many chickens his family raised, whether there was netting over the coop, and whether he's tried to raise chickens since. You may disprove his points without even trying.

ANDY'S ANECDOTE

AFTER ONE CITY council meeting, a woman against backyard poultry approached me to ask what chicken owners did with their chicken poop. I explained that some people add it to their gardens, compost bins, or around flowers and flowerbeds and that others scoop it up, place it into a trash bag, and put it into the garbage. At that point her eyes widened. "You put chicken poop in the garbage?" she asked. I responded by noting that we put human and other animal refuse into the garbage so why not that from chickens? She soon realized her question was moot and walked away. A little logic goes a long way.

PREP WORK:
GETTING YOUR HOME AND YARD READY

YOU'VE LEARNED THAT you can legally keep chickens in your backyard. Now, before you bring home your chicks or eggs—whether your first or fiftieth set—you need to prepare. Gather all your equipment and keep it nearby in a handy space. Stock the necessary food, water, and vaccines. Do a once-over of your yard and tools to make sure everything's in good repair before your birds arrive. To really get ahead of the curve, plan for your coop to include water collection; you can accumulate and store rainwater runoff for watering outdoor runs or a nearby garden or to fill coop waterers.

In chapters 5 and 6, we discuss the specifics of chick incubation and brooders, respectively. In chapter 7, we provide details about setting up a coop. This chapter is intended to get your wheels turning in the right direction to make your home chick- and chicken-friendly.

SETTING UP SPACES IN THE HOME

Some of the spaces you'll need to consider for your birds as follows:

* A place to incubate/hatch eggs
* The brooder
* The coop
* A quarantine pen—essential to plan ahead

Know the ins and outs of a safe and healthy environment
for your flock and have a solid plan in place.

Hatching your chicks in a home setting requires planning ahead to ensure an ideal location with all the necessary equipment and environmental controls.

Incubation/Hatching Area

Hatching chicks at home requires a special area for this activity. If you value your eggs and the chicks they will bear, look critically at all potential spaces. An ideal incubator area is well ventilated but draft-free, a place away from high traffic, and one where you can regulate temperature. A spare bedroom or bathroom, for example, works well. If you worry you might forget about incubation-related duties because the incubator is out of sight, hang reminder notes in high-traffic areas or some place you look frequently, such as your bathroom mirror.

You also need a sturdy table on which to place the incubator. This is especially important if children will likely lean on the table to examine the incubating eggs. Also, within your room of choice, do not place the incubator or the table directly above or below an air vent, as you can inadvertently cause incubator temperatures to fluctuate.

Brooder Space

Once you have chicks, whether you hatch them at home or buy them, you need a room suitable for setting up a brooder (described further in chapter 6). This should be draft-free and separate from the room where you incubate your eggs (if you are incubating your own). Again, high-traffic areas near doors that lead outside create drafts, so select a spare room, if possible. Because chicks create messes, try to find a room with a linoleum floor for easier cleanup.

As tempting as it may be, it is not a good idea to set up a brooder in a child's bedroom. Children tend not to have good hand-washing habits, increasing the potential for them to pick up *Salmonellosis* or other illnesses from the chickens. (Chickens can also get something from children, too!)

A garden chicken run and coop

A shady location is a plus
in hotter climates.

The Coop

Where to put a coop on your property is a personal decision, but here are some tips to help you decide.

- Take advantage of trees in your yard for shade during the hottest times of day. This will cut down on coop-cooling costs during hot months.
- Also, consider carefully when deciding between mobile and stationary coops (we get into this in much more detail in chapter 7). You can move a mobile coop during winter days to parts of the yard that receive extra sunshine.
- If you opt for a stationary coop, locate it near a water source for ease of filling and cleaning waterers.
- If possible, locate your coop near an electrical source for running a fan during the summer and heat lamps or electric water-base heaters during the winter.

Remember to incorporate local zoning requirements. If you live within the limits of a city, there may be an extra layer of complication to this process. Cities each use different language to refer to chickens and the coop (refer to chapter 3 for more about laws). Look at the laws referring to animals and also to accessory structures. Some cities consider a coop an accessory structure much like a shed in your backyard. For the coop, that could mean required distances from a property line or a fence. Height may also be a factor. The bottom line is to do your homework before making a purchase.

Water Sources

A fresh, running water supply is essential.
Base heaters are available to ensure drinking water
is always available in colder weather.

Frozen waterers present a serious problem to owners
of small flocks. Electric water bases, when used with
metal waterers, work well and last several years.
Plastic waterers tend to crack in the winter due to the
daily freezing and thawing associated with low
outdoor temperatures.

Quarantine Pen

Few chicken owners think about a quarantine or sick pen
until an emergency arises. This is a place to relocate
sick or injured birds as they heal. Think about this pen's
location early, as you prepare for chicks. Where will it
sit in relation to the regular coop?

Locate the sick pen as far as physically possible from the
regular coop. Is there a far side of your house where no
one goes frequently? This could work. Also think about
airflow on your property. If possible, situate the quarantine pen downwind from your regular coop in case a flock
member picks up an airborne pathogen. You can use
something as simple as a dog crate for this pen, as long
as it is secure from predators and the elements:
something that won't overheat during the summer
or freeze during the winter.

Many people are forced to stick the sick pen in the
garage or in another high-traffic area because they did
not plan ahead. This is not preferable. The goal of a
quarantine pen is to keep the organism residing in the
bird far away from the rest of the flock. If you keep your
feed, boots, or barn clothing in the garage—in the same
place as the sick pen—then you may track the organism
right back to the rest of your flock. Find an area with low
foot traffic that provides quiet in which the chicken's
body can repair itself, such as the following:

> ✦ A spare bedroom
> ✦ A side shed near the house
> ✦ A dog kennel nearby
> ✦ A nook somewhere near the barn

A quiet, isolated sick pen is important for the ailing bird as well as her flock members.

Plan for all eventualities. Keep separate equipment in the sick area because dirty equipment shared around the farm or yard also can spread disease (see the "Biosecurity" section of chapter 10). Mark equipment designated for the quarantine area with colored tape (such as an eye-catching red). Remember, to keep a flock happy and healthy, you need to plan ahead for potential problems.

Cat curiosity could be injurious to your chicks, so make sure chicken wire
is in place and repair damage immediately.

PREPARING FAMILY PETS

Before your chickens or chicks come home, consider any
family pets already at your residence. Some dogs and
cats have strong predatory instincts and may exhibit
extremely high prey drive toward your chickens. In other
words, they'll attack anything that looks like prey. If you
are unsure whether your pets have this tendency, try
a preliminary introduction with training in mind. For
example, if you have friends or family with chickens,
borrow and bring home a few feathers for the dog or
cat to smell. Don't make a big deal about this new item.
Rather treat it like something completely normal and
expected for the pet.

Cats

Cats usually do not like traveling to new places, period,
so we don't recommend bringing them to a farm for a
through-the-fence introduction to chickens. Cats tend
to pose the greatest risk when you brood chicks inside
the home. A housecat may like nothing more than to
"play" with your new baby chicks. Obviously, that's
a no-no. Place a window screen over the brooder top
to keep kitty out.

A dog can be a chicken's best friend against predators once he finds common ground and bonds with the flock.

Dogs

Dogs, on the other hand, tend to handle traveling well. Start with an on-farm, through-the-fence introduction to chickens and watch the dog closely. Immediately correct unacceptable behaviors such as excitement, whining, barking, or focused, intense gazes toward the chickens. Make the dog lie down or face away from the chickens until calm. Keep the canine's attention on you as much as possible. And remember that your level of tenseness and excitement transfers right down the leash to the dog. Act nonchalant but in charge. If you have multiple dogs, conduct separate introductions so the behavior of one dog does not influence the other.

Not everyone is a professional dog trainer, but you usually know if you have a good handle on your pet's behavior. If you already experience difficulties training animals, perhaps chickens are not for you. Remember in some hierarchies, they fall into the "prey" category; be sure to predator-proof the coop (see the "Protection Against Predators" section of chapter 7). Even if you feel you have enough distance between your pets or sufficient control over them, always plan for the worst-case scenario. There will be a day when someone accidentally leaves a gate ajar or unlocked and your pet gets out and goes after the chickens. Plan your coop well—site it away from your dog kennel, for example—to prevent against this eventuality and others like it.

CONSIDER WHAT'S IN YOUR YARD

The yard itself can present all sorts of challenges to consider before bringing home chicks or setting up shop for a flock. A chicken spends about 60 percent of her day exploring her surroundings, searching for food or other interesting objects. While she performs this exploratory behavior, her attention is divided, making it even more important for the yard to be clear of dangers and safe from predators.

Birdfeeders

Birdfeeders can attract disease-carrying wild birds (see the "Biosecurity" section in chapter 10) so you must view them as "dirty." Birdfeeders carry disease.

If you already own a birdfeeder, you really have two options: Get rid of it altogether or relocate it as far away from the coop as possible. Place it in the front yard, for example. If you want to keep the feeder, make it the absolute last stop in your daily routine, after you've finished with your chickens. Also, fill and clean your feeder at an area or sink separate from the one you use to fill and clean your chicken's equipment. Now that's good biosecurity!

Beneficial and Poisonous Plants

The list of vegetables and fruits chickens will eat is long and includes almost anything you eat. Many of the garden plants to which you expose your chickens are fine for them to peck at, but they shouldn't eat the seedlings. Rather than cause yourself woe, wait until your garden matures before letting in your bug-eating, garden-tilling flock. (Even then, you may still need to block off areas of the garden you want your hens to avoid.)

Birdfeeders can attract diseased birds as well as a variety of predators. Place yours far enough away from your chicken's grazing area to be safe.

NATURAL INSECTICIDES

In terms of plants and trees to place around—not in—your coop, consider these, which have natural insecticidal properties (find out more information about them from your cooperative extension office).

These plants aren't good for your chickens to eat, but they may prevent the forward-march of mites or lice toward your flock.

Chrysanthemums
(*Chrysanthemum morifolium*)

Lavender
(*Lavandula angustifolia*)

Comfrey
(*Symphytum officinale*)

Bay laurel
(*Laurus nobilis*)

Tansy
(*Tanacetum vulgare*)

Neem tree
(*Azadirachta indica*)

Artemisia or wormwood
(*Artemisia vulgaris*)

Foxglove
(*Digitalis purpurea*)

Nightshade
(*Atropina belladonna*)

POISONOUS PLANTS

Finally, let's look at poisonous plants. Because they're often difficult for the average homeowner to identify, we recommend consulting your local extension agent or master gardener. Keep in mind that seasonal seeds, pods, or acorns—which you may not see at the time of year when you're readying your coop—can also pose risk, so plan ahead when you are choosing a site for your coop.

The toxicity of these plants may be restricted to their seed, a part of the plant, or a stage of growth, so read up before landscaping with some of these plants. It's not necessarily the first bite you should worry about, but rather chickens consuming large quantities over time.

Shown here are some poisonous plants to remove from your yard.

Oleander
(*Nerium oleander L.*)

Bladderpod/bagpod
(*Isomeris arborea*)

Corn cockle
(*Agrostemma githago L.*)

Black locust
(*Robinia pseudoacacia L.*)

Castor bean
(*Ricinus communis*)

Crown vetch
(*Coronilla varia L.*)

THE CHICKEN WHISPERER'S GUIDE TO KEEPING CHICKENS

Death camas
(Toxicoscordion venenosum S.)

Poison hemlock
(Conium maculatum)

Water hemlock/cowbane
(Cicuta maculata)

Jimsonweed/thorn apple
(Datura stramonium)

Pokeberry
(Phytolacca Americana)

Yew
(Taxus baccata)

Milkweed
(Asclepias syriaca)

Rattlebox
(Crotalaria verrucosa)

READING THE YARD

Your yard's almost ready for chickens. Here are a few other factors to consider.

Fencing for Pasture Rotation

If building a coop for the first time—and you don't opt for a mobile one—carefully consider the outdoor area. (Chapter 7 discusses coops in detail.) The most up-to-date thinking in chicken-coop planning has moved away from a single outdoor run toward a coop with four runs. The more outdoor runs or pastures available to the chickens, the better it is for your flock because one spot doesn't become a denuded, muddy mess. With only one run, a mess is guaranteed.

Plant different grasses and forages in each run. For example, in one, grow shade cover such as sunflowers or corn among shorter grasses. This will allow your chickens a shady run in the height of summer that is also edible later in the season. In different pens, spread out tall fescue, Bermuda grass, and alfalfa, as well as red, white, and ladino clovers. Chickens love clover and can quickly decimate a pen filled with it; plant clover in conjunction with another grass to keep the pen usable for longer. The notion of maintaining multiple runs will require you learn more about grasses or forages than if you had a single run, but it is one of the most modern concepts in coop planning.

Exclusion of predators is important for these outdoor runs, so bury the bottom of chicken wire or use electric fencing. Any aviary netting over the top of the outdoor runs should be taller than the tallest family member who must enter.

A Rock Border

When you design your outdoor runs, keep in mind the daily activities that take place in your yard: lawn/yard maintenance, sprinklers spraying, children playing, or people eating on the patio. Each of these can stress your flock. Mowing the lawn, for example, means bringing out a loud, scary machine. Build a 12- to 18-inch (30 to 45 cm) rock or dirt border around the coop and runs. This prevents the machine from getting close to the pen's edge, therefore reducing fear in the birds. To make mowing a positive experience for your flock, make your first pass with the lawnmower one that blows clippings directly into the pen for the chickens to eat. Or gather the clippings in a sweeper and dump them into the pen for a tasty treat for the chickens. Don't do this if you treat your lawn with harsh chemicals as it may harm the birds.

Do your best to keep the grass cut and the weeds low around the coop. Having a border around the coop means less weed removal because fewer weeds will grow in that spot. Also, low grass prevents rodents from building a highway from the weeds that can grow up next to coops.

Sprinklers

Direct sprinkler systems away from the feed containers and pens because the water can quickly make the pen muddy and the feed moldy. If you wish for your yard sprinklers to also water the runs, consider running them at night when the flock is inside and not likely to get soaked. Chickens tend to stay inside on rainy days because they dislike getting wet.

Children can hardly contain their enthusiasm when they see an active flock of hens and chicks, but they need to be taught the proper way to handle poultry.

Human Activity

Children running around playing can scare a new flock. The birds must get used to balls, loud toys, and shouting by children. Teach the younger members of your family to treat the coop and its flock with respect. Also, consider the smoke of a barbeque and activity of dinner on the patio. Chickens do not like smoke that drifts into and lingers in the coop. Also, chickens beg for treats just like any other pet, so unless you want to hear their cackle and endure hens' stares throughout your dinner, erect a visual barrier between the chickens and where you cook and eat.

This chapter provided you a few details for your consideration, pointers meant to help you give your chickens an excellent overall experience. In the coming pages, we turn to chick incubation, brooders, and coops.

iNCUBATiON: SO YOU'RE EGG-SPECTiNG?

A WEEK DOES NOT PASS without one of us telling someone about the amazing experience of incubating and hatching chicks. We have hatched out many times, and the twenty-fifth time is just as exciting as the first! It only takes twenty one days for a chick to hatch. It's a great project for young and old alike. As we mentioned in "Setting Up Spaces in the Home" section of chapter 4, you must find a well-ventilated, draft-free, low-traffic area such as a spare bedroom or bathroom to incubate and hatch your chicks. And don't forget about a sturdy table on which to place an incubator.

INCUBATION MATERIALS

To create a successful hatching environment, you need the following:

- ⚡ An incubator
- ⚡ Water
- ⚡ A thermometer
- ⚡ A hygrometer
- ⚡ An egg turner

You will never grow tired of the remarkable experience of watching chicks make their way into the world.

Incubator

Incubators come in all shapes and sizes with two standard styles: the tabletop version, typically made from Styrofoam, and the cabinet version, which holds eggs on built-in shelves. Most home users choose the tabletop style because they're more convenient and less expensive than cabinet style. Incubators also range in costs so they fit in any budget. With a little know-how, you can even build your own. An incubator contains two thermometers: a dry-bulb (regular thermometer) and wet-bulb (relative humidity) thermometer.

Thermometer (Dry-Bulb)

For a successful hatch, you need to closely watch your incubator's temperature. A chick embryo begins to grow at 86°F (30°C), but for appropriate development to progress, the incubator must maintain a temperature of at least 99.5° (37.5°C). (Believe it or not, some avid hatchers have identified a temperature down to one hundredth of a degree above or below that.) From our many incubation experiences, we have found that 100°F (38°C) also results in a successful experience. Adjust the incubator's temperature using the built-in thermostat and heating element.

The tabletop model is an alternative to the more-expensive cabinet model, above, which offers an accurate digital thermostat with LCD display of termperature and humidity.

A thermometer is used to monitor the temperature of your incubator.

A hygrometer is a tool scientists use to measure relative humidity.

Place the thermometer inside the incubator to ensure it starts at and maintains a correct temperature consistently throughout the entire incubation period. Even with an incubator that self-regulates inside temperature, check it at the outset and daily throughout the incubation process, to ensure the incubator consistently holds that temperature. If you have experienced difficult hatches in the past, you may opt to use two thermometers. By placing two inside the incubator, you can compare their readings to confirm accuracy.

Hygrometer (Wet-Bulb)

The hygrometer, an instrument that measures humidity level or moisture content in the air, is just as important as the thermometer in creating a successful hatch. Hygrometers are easy to find, both online or at the local cigar or hardware store. For the first seventeen days, the incubator needs to maintain 45 to 55 percent humidity. To reach this, it should read 82°F (28°C) to 84°F (29°C). On the eighteenth day, increase the humidity level to 55 to 65 percent. The hygrometer should read 86°F (30°C) to 88°F (31°C). Keep it at this level until day twenty-one.

Maintaining Humidity

If you have difficulty creating an environment with enough humidity, add a damp sponge to the bottom of your incubator. This increases the moist surface area and allows more water to mix in with the incubator's air, which in turn increases the relative humidity. Try this if you consistently have problems with chicks sticking to the inside of their shells as they hatch or if you experience egg loss.

ANDY'S ANECDOTE

EXTREME OUTSIDE temperatures can influence the inside temperature of an incubator. For one of the first hatches I attempted, I placed the incubator on the kitchen counter to easily and regularly check the temperature. What I didn't realize, however, is that the direct sunlight from the kitchen window increased the incubator's temperature to over 115°F (46°C), ruining the eggs. It was a lesson I did not have to learn twice.

Create a system where you mark your eggs to track their rotation.

Xs and Os

If you are not usiing an automated egg turner,

you can keep track of your rotation schedule by using a pencil or marker to lightly draw an "X" on one side of the egg and an "O" on the opposite side. On your calendar, mark on which side you started so you know when each letter should face up and on which you will end. This can be particularly important if the eggs get bumped or accidentally rotate out of place.

Egg Turner

For the first eighteen days—no more, no less—the eggs inside the incubator must be rotated a minimum of three times daily. You can do this—carefully—by hand, but if you do, try to complete the task as quickly as possible to prevent too much heat from escaping from the incubator. Also, make sure to wash your hands *and* use hand sanitizer before and after handling the eggs.

If you don't want to turn eggs by hand, try an automatic egg turner. It's certainly not required, but it's not too expensive and sure makes the process easier, particularly on busy days when getting to the incubator is a challenge. An automatic turner, which constantly rotates the eggs so slowly that it's hard to see with the naked eye, plays the role a hen does in nature, naturally moving the egg around in the nest. The movement keeps the embryo centered inside the egg so it doesn't stick to the side membranes.

We highly recommend an automatic egg turner for hatching chicks in classrooms. Though turning eggs by hand may be a fun activity for children, it's sometimes inconvenient. Plus, hatching eggs in a classroom without an automatic turner means packing up the incubator, eggs, and equipment every weekend for three weeks so eggs can be turned at home over the weekend.

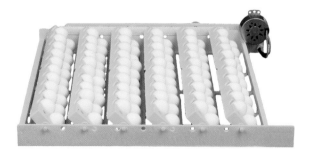

An automated egg turner

GETTING THE EGGS

To hatch chicks at home, you need fertile eggs, and unless you have a rooster, your hens don't produce eggs that will hatch. If you do have a rooster, you should get fertilized eggs. For about ten days after a rooster breeds a hen, she lays fertilized eggs. To determine whether embryos are growing, approximately seven to ten days post-incubation, hold a strong light to the outside of the shell (as shown below and on page 73). You'll actually be able to see a developing embryo. This is called candling. Throw away those that are not developing; they can develop mold or bacteria or even explode.

If you don't have a rooster, there are many ways to obtain fertile eggs. Do you know other flock owners who could provide them to you? What about a local farmer or hatchery from where you can purchase them? You also can purchase fertile hatching eggs online from poultry breeders or businesses such as www.mypetchicken.com. Be aware that fertilized eggs sometimes cost more than day-old chicks, particularly for rare breeds.

If your home flock produces fertile eggs, collect them daily from your hens and store them in a clean egg carton at a temperature between 50°F (10°C) and 60°F (16°C)—the average refrigerator is too cold—and at 75 percent relative humidity for no longer than two weeks. The fertility of eggs stored for longer than a week decreases significantly. Do not wash the shell of eggs destined for incubation. Washing removes the shell's bloom, which seals its pores and prevents bacteria from entering. The bloom eventually wears off during incubation, allowing air to enter and leave the egg for the benefit of the chick. Instead of washing the eggs, buff off stains or dirt with a piece of high-grit sandpaper.

Once you collect the eggs you want to hatch, place them all into the incubator all at once. Why? To follow the lead of hens in nature; these animals don't sit on their nests until after they've laid clutches containing several eggs. For your purposes, as eggs set in the incubator, embryos begin to grow. Twenty-one days after you place an egg into the incubator, a chick will hatch. If you do this as you collect the eggs, all your chicks won't hatch together. Chicks typically prefer being brooded in groups, and the convenience of having many chicks born at the same time means opening the incubator minimally and caring for the babies together. (Don't worry about late bloomers; chicks talk to each other to speed up slow hatchers.)

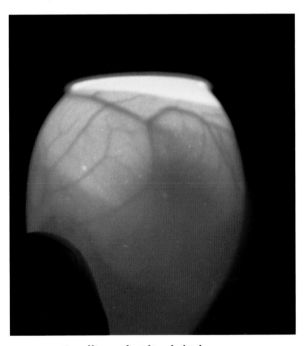

Candling a fertilized chicken egg

Washing the Right Way

Only wash eggs bound for incubation when they are excessively dirty. Here's the right way to do it so you do not push bacteria from the shell's outside in, potentially killing the embryo.

① Warm running water to a temperature hotter than that of the egg, somewhere between 110°F (43°C) and 120°F (49°C). Reaching this temperature ensures that the osmotic pressure pushes any bacteria in the pores outside of the egg.

② Dip a soft cloth in the warm water and wipe the outside of the shell. Do not immerse the egg in the water. For stubborn dirt, add a mild detergent to the cloth. (Specialized egg-washing detergents are available, but a gentle household soap also does the trick.) After you wipe the egg with the water-soap mixture, rinse it under warm running water.

③ Dry the egg using a clean towel.

Putting an egg in water colder than its temperature pushes bacteria from the shell's outside inward because you are cooling the egg. This defeats the purpose of washing and may cause an egg with sufficient bacterial growth to explode in the incubator. The explosion has the potential to spread bacteria to all other incubating eggs, possibly destroying a larger number of embryos.

READY TO INCUBATE

Have in mind a number of eggs you'd like to hatch. Once you near that amount, plug in the incubator two days prior to setting your eggs to ensure that everything works properly. Place a thermometer inside to test the temperature, which should range from 99.5°F (37.5°C) to 100°F (38°C). Add water to the incubator's tray and use a hygrometer to maintain the appropriate humidity level (45 to 55 percent until the eighteenth day, 55 to 65 percent after that).

Also, remember that eggs must be turned a minimum of three times a day. Egg turners do influence the incubator's inside temperature, so install the turner two days prior to placing the eggs inside. This will let you adjust the temperature to compensate for the heat of the egg turner's motor.

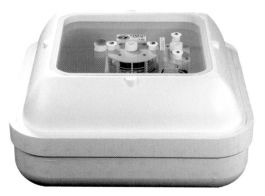

Be sure to set up your incubator prior to beginning the hatching process so all systems are in order.

Use a candler to check the progress of an egg's development.

Keep It Closed

For the best hatch percentage, open the incubator only when absolutely necessary so you can trap hot air inside. Most incubators come with two or more air holes on top. Cover all but one until the chicks hatch so oxygen is present but heat stays in. Uncover all the holes once the chicks hatch.

With your incubator ready, place the eggs inside. If turning them by hand, implement a system to track rotations (see the sidebar, "Xs and Os", on page 70 for a suggestion). If using an automatic egg turner, place the eggs in with the small end pointing down. On the eighteenth day, stop turning the eggs, remove the egg turner, monitor, and adjust the inside temperature. Lay the eggs flat on the floor of the incubator. Your chicks are almost here!

Eggs sitting small end down

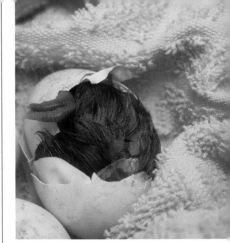

Chick has already "pipped" (pecked a hole in the shell with its beak) and is now working to enlarge the hole.

The process of enlarging the hole by pecking around the entire shell, splitting it in two, is sometimes called "pipping" or "zipping."

The chick is starting to push its way out of the shell.

THE HATCHING PROCESS

On day eighteen of incubation, after you remove the automatic turner (if you used one), lay the eggs on their sides in the incubator. On this day, the chicks begin their two-part pipping process.

First, during internal pipping, chicks break through the air cells inside the eggs and take their first breaths. (Sometimes, if you are really lucky, when you hold an egg up to your ear, you can hear the chick pecking or chirping.) During this stage, the chicks communicate, encouraging each other to speed up hatching so they are born simultaneously. In nature, the mother hen starts talking to the chicks on this day as well, encouraging them to hatch. Studies have been done during which eggs are hatched singly, in groups, and under a hen. The rate at which chicks hatch speeds up when they communicate with each other.

The second stage is external pipping, during which the chick makes its first break through the eggshell. This crack is small and difficult to see. The chick is using its egg tooth (as seen in the photos above) to crack open the shell. The egg tooth falls off a few days after hatching.

Do not help the chicks out of their shells. They must use their special hatching muscles to get out. Also, by turning inside the eggshell, the chicks pull in blood from the circulatory system and the yolk. Without these essential components, chicks may bleed uncontrollably. There is one instance, however, when you should help a chick: If it becomes stuck to the inside of its shell due to a dried remnant of umbilicus. When this happens, free the chick by snipping through the umbilicus piece with scissors.

As chicks hatch, they are often attracted to movement. Many first-time chicken hatchers incorrectly believe that a chick that runs over to an egg still hatching is trying to help its fellow hatchling. This is not the case. The chick is simply responding to the movement.

AFTER INCUBATION

Once the chicks are out, you'll focus much of your attention on their care. However, remember that every incubator, no matter the size, requires a thorough cleaning after every batch of chicks moves to the brooder. It may be boring work, but it is crucial. Bacteria and fungi love the little bits of shell, blood, and feces your

The chick is using its feet to push the bottom half of the shell away.

She's hatched! Chicks come out of the egg wet, but when they dry, they look more like the fluff balls you've been expecting.

chicks leave behind. If you ever wish to hatch eggs again, washing is a must.

Here's how you do it:

① Sweep up. This includes removing all fluff (a chick's first feathers, technically called natal down) and shell bits.

② Dump out any water that remains in the incubator and remove sponges. (You will have already removed the automatic turner, if you had one, but do not forget to clean it as well.)

③ Take apart the incubator and wipe down all surfaces with warm, soapy water. Avoid coming in contact with electrical components. Rinse the sponges by squeezing them out in hot, soapy water.

④ Wash the sponges and all surfaces to get rid of any soap. Heat the wet sponges in the microwave for fifteen seconds. They will come out hot so let them cool before touching.

⑤ Dry everything completely and then use a disinfectant. This will rid your incubator of almost all remaining harmful organisms. A disinfectant can be a simple bleach solution of ¼ cup (60 ml) of bleach in 17 cups (4 L) of water. For Styrofoam incubators, do not use bleach because the material may break down if left in place for too long. Instead try products such as TEK-TROL, Nolvasan, 1-Stroke Environ, or Oxine. Always apply disinfectants according to the label directions. The amount of contact time needed for the surface varies with each disinfectant. To apply the disinfectant, wear gloves, spray with a bottle, and wipe with a clean sponge.

⑥ Use a clean sponge to wipe off the disinfectant after the prescribed contact time and then let the incubator dry thoroughly.

⑦ Put everything back together and test your incubator by turning it on to make sure it works.

THE LIVING INCUBATOR

Are you considering hatching eggs using hen power rather than an incubator? Make an educated decision after weighing the pros and cons. Setting refers to a hen's desire to incubate eggs. A hen willing to sit on and hatch eggs is said to have gone broody.

Pros of the Hen

One benefit of having a hen go broody is a lower electric bill. It is, indeed, a greener option than an electric incubator.

Also, it is nice for children (and adults) to see a hen brooding eggs and chicks. As with any new chick, be aware that you will not know whether your chicks are pullets (females) or cockerels (male) until they are much older. If you end up with male birds that aren't permitted on your property—the law in some places—have a backup plan in mind (for example, a place to which you can sell them).

Cons of the Hen

Some hens are overly broody. That means they choose to sit on eggs rather than to eat, drink, or leave the nest. These hens require constant monitoring or else they may die on the nest. Also, some hens go broody without any eggs upon which to sit. Broody hens undergo a change to their bodies and hormones so that they stop laying eggs. If you want eggs from a hen, discourage broody behavior by dunking her in ice-cold water for one minute, then letting her dry in a warm area. For a few days following the ice-water bath, keep her out of the area where she wants to sit. She may be "as mad as a wet hen," but you need to break her broody habit if you want eggs.

Secondly, you must check hens daily for chicks, in case they come out in a staggered hatch. If you wish to ensure the health of your flock against the highly contagious viral Marek's disease (discussed in detail in chapter 10), keep several bottles of vaccine for this disease on hand. Chicks must be vaccinated during their first twenty-four hours of life. Plus, once mixed, the vaccine only stays alive for four hours.

Third, you need a hen with a strong setting instinct, something not all hens possess. Take advantage of a hen willing to sit on eggs, but know that such timing may not correspond with when you wish to hatch eggs. Though hens of any breed are capable of going broody, chicken breeds with strong setting instincts include Silkies and Cochins.

Finally, mother hens do an excellent job protecting their chicks, but they need a safe, secluded area separate area from the rest of the flock for the first couple of weeks after the chicks hatch. The chicks also need a separate feeder. The hen may continue eating a laying-hen pellet or crumble diet; however because she is brooding, she is not laying, so she can eat the same food as her chicks. In fact, layer pellets are harmful for chicks because of their high calcium content.

Mother hens will teach their chicks how to eat and drink. You may need to monitor the lessons in case they do not catch on right away. Without a doubt, your interaction with the hen will raise her defenses, so be forewarned that you may become henpecked!

Some breeds have a stronger constitution when it comes to setting, such as the Cochin or Silkie. Leghorns, particularly single-comb White Leghorns, are not known to have a strong instinct to set. Research your breed type in this aspect before deciding whether you should invest in an incubator.

It is amazing how quickly your chicks develop after the hatching process. It's time to prepare for the next batch!

Not all hens have a strong setting instinct. Make sure you find a hen willing to set if you prefer a living incubator.

THE ART OF BROODiNG: PREPARiNG THE NURSERY FOR YOUR NEW ARRiVALS

IF CHICKS ARE IN YOUR FUTURE, regardless of whether you hatched them using an incubator or ordered them from a hatchery, you will need a brooder for your new arrivals.

A brooder, your new chicks' home for their first few weeks of life, is a box or pen that provides them a warm and dry place to live and a fresh and continuous supply of food and water, as well as protection from predators and a potentially harsh outside environment. It does what a mother hen would do in nature.

Your chicks live in the brooder—which is frequently indoors but always somewhere temperature-controlled—until they fully feather, usually between six and eight weeks of life. Actual length of time, however, depends on outside temperatures and the season; in January, they may need up to twelve weeks, but during summertime, they may need just four.

A brooder consists of several parts:

- ➤ Container
- ➤ Heat source
- ➤ Waterer
- ➤ Feeder
- ➤ Bedding
- ➤ Thermometer

⟦ A baby chick taking a drink from a commercial brooder trough ⟧

A brooder box does not have to be fancy. You have three options for obtaining a brooder: Purchase a commercial one, buy a kit that includes some brooder parts and requires you to provide some, or make your own homemade brooder out of wood, a kiddie pool, a plastic or rubber storage bin, a cardboard box, and so on. We get into more detail about this later.

First, determine how big a brooder you need based on how many chicks you expect. Remember, chicks don't stay small for long and they need more space as they grow. Really, you can use anything for a brooder that has tall enough sides and ample ground space square footage.

SELECTING A BROODER

Because we all have different expectations and preconceptions for brooders, different types work for different people. When picking out your brooder, consider one that works well for you, your schedule, and your chicks.

ANDY'S ANECDOTE

ONE DAY, I CALLED MY WIFE with what I believed was a revolutionary idea. I suggested that we use a kiddie pool as a brooder for the twenty-five chicks we would soon receive. We rushed to the store and bought a cheap plastic kiddie pool and brought it home, eager to set it up. Turns out, it was one of the worst ideas I've ever had.

It worked fine for the first week. Then the chicks started flying out to explore the garage. We purchased netting and lashed it down around the kiddie pool, but we soon found chicks marooned between the pool's outside walls and the netting. We haven't used this brooder option since. I tell this story not to discourage, but rather to show that people have mixed results with every brooder option. Many people love the kiddie pool idea. If it appeals to you, give it a try; what doesn't work for one chicken owner may work perfectly for another.

A commercial brooder box

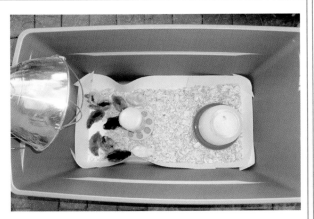

A completely functioning homemade brooder
made from a rubber/plastic bin

A homemade cardboard brooder box

Essentials: The Container

Many household items can pass for a brooder, for example, an old bathtub, an out-of-use child's playpen, or an old animal trough. A favorite homemade brooder, with which we've started countless new chicken owners, is a simple 45-gallon (170 L) storage bin purchased from any big-box store or major retailer. It has many qualities that suit it for the task:

- → Tall sides make chick escapes unlikely.
- → An elongated shape allows plenty of room for growing chicks to move toward and away from the heat source.
- → Sturdy sides and handles offer a perfect place to clamp a heat lamp.
- → Plastic or rubber material makes it easy to clean with a household garden hose.
- → Wheels, found on some models, make it easily moveable.

Cardboard boxes can work, but with other free or inexpensive alternatives available, you may want to stay away from this potential fire hazard. Styrofoam coolers provide ample space for a few chicks, but these babies love to scratch and peck, often doing so to the small Styrofoam balls which make up the cooler's walls. For that reason, we discourage their use as well.

Commercial brooder boxes are fantastic if you have the budget to purchase one. With water troughs, feed troughs, built-in heaters, and removable floors for cleaning, they make for an easy and pleasant brooding experience.

Essentials: The Heat Source

Now that you've selected your brooder, you need its corresponding pieces. First, the heat source.

Chicks get cold quickly because they haven't yet feathered out (they just have down) and therefore haven't bulked up their internal heat source. There's no mother hen to keep them warm, so as chicken owner, it's your job. Do this by hanging a light bulb into the brooder to provide heat. Just make sure to choose the appropriate wattage, which we discuss shortly, and to hang it in such a way that leaves 18 inches (45 cm) between the bulb and the chicks. Otherwise, the heat bulb can become dangerous. We recommend a brooder lamp or shop lamp in conjunction with a heat lamp bulb as seen in the photos at right.

There are two basic lamp types. One has a direct wire from the outlet to a ceramic base where the bulb is inserted. The connection of the plug to electricity alone turns this lamp on and off. The second lamp has a wire that connects from the outlet to a plastic base, which also has an on/off switch. For safety reasons, we discourage using cheaper plastic lamps. We've seen them melt due to high heat from constant bulb use in the brooder.

BROODER TEMPERATURE

All chicks need access to a spot within the brooder that is 95°F (35°C) for the first week and 5°F (about 3°C) less each week thereafter until the brooder temperature gets down to no lower than 70°F (21°C). Attach a thermometer to the section of the brooder at the chicks' height level to confirm that it's at the appropriate temperature. If it hangs on the side of the brooder, it may be too close to the heat lamp to provide an accurate reading.

A direct wire brooder heat lamp
with ceramic base

A red heat lamp bulb for use
with the brooder heat lamp

A clear heat lamp bulb for use
with the brooder heat lamp

If you pay close attention to your chicks, however, you do not truly need a thermometer in your brooder. If there is a draft in the brooder, your chicks will chirp loudly as they huddle in the corner that provides them the greatest protection. Through their actions and their peeps the chicks will let you know whether to increase or reduce the temperature. Try this: After adding heat to the brooder, watch how the chicks react. If all the chicks are loud, huddled close to the heat source, or lethargic, they are too cold. Up the temperature either by lowering the lamp or swapping in a higher-wattage bulb. If this does not solve the problem, consider the fact that you may have a draft. If the chicks stay far away from the light, chirp loudly, or pant, they are most likely too hot. A reaction to a heat source that's too hot may take time. After you increase the temperature, check the chicks every fifteen minutes to be sure they do not overheat. If the brooder's too hot, correct this problem by lifting the heat source above the chicks or by putting in a lower-wattage bulb.

The chicks do not need a uniform temperature through-out the entire brooder. It is best to keep one area of the brooder warm, but not too hot, and keep other areas cooler so the chicks can regulate their own comfort levels. We have seen more chicks die from overheating than from being too cold because chicken owners put the chicks in a small box with a heat source that's too close. This is why the brooder needs to have an oblong shape or wide base that allows the chicks to get away from the heat source.

Whether you choose to clamp, hang, or set the heat lamp on the brooder, make sure it is secure with no chance of falling from its berth. Remember, all heat sources are fire hazards and should be treated with caution.

Happy Chicks

How do you know whether your chicks are content? Happy chicks spread out in the brooder and move around freely, scratching, eating and drinking, and cuddling under the light to sleep. They quietly chirp happy chirps.

LIGHT BULBS

There are two widely used types of bulbs—red and white light bulbs. Both give off the same wattage of heat. The only difference (literally) is what color light they emit. Some research shows that a red bulb discourages chicks from pecking at each other. While that may be a researched truth, in our experience, chicks with ample space peck little, if at all. And just like people, chicks crammed into a small space get irritated with each other, which may translate into pecking.

Several factors determine the wattage appropriate for the bulb. A 125-watt bulb usually works for chicks in a brooder housed in a controlled-temperature environ-ment such as a basement. Consider a 250-watt bulb if the environment outside the brooder is cool, such as a garage during wintertime. And know that you can always change the bulb. We recommend swapping bulbs as the temperature changes with each season.

A plastic one-quart (1 L) feeder with plastic base for use inside the brooder

A plastic one-gallon (4 L) waterer for use inside the brooder

Essentials: Waterers and Feeders

Feeders and waterers come in all different sizes. The number of chicks you need to feed and provide water and the size of your brooder box will determine your choice. Most commercial brooders come with their own feeding and watering systems, like the model in the photo on page 81, which shows trays on the brooder's outside. There are also plastic or metal feeders and waterers in a range of sizes that you can purchase from your local feed-and-seed store for your homemade brooder. Provide 1 inch (2.5 cm) of feeder space per chick during the first week and 2 inches (5 cm) of feeder space until the end of brooding at six weeks old.

You can make homemade feeders out of bowls or even egg cartons and homemade waterers out of plastic bowls, plastic bottle bottoms, or Mason jar lids. The drawbacks to homemade feeders and waterers are that chicks will climb into the feed and water, excrete into them, sleep in them, and ultimately knock them over, spilling the contents. You can certainly hot glue the homemade feeder or waterer to some cardboard so the chicks can't turn it over, but it is best to use the feeders and waterers specifically designed for chicks found at a feed-and-seed store or online.

BECAUSE CHICKS ARE ATTRACTED TO SHINY THINGS, FILLING THE HOMEMADE OR STORE-BOUGHT WATERER TRAY WITH MARBLES ENCOURAGES THE CHICKS TO PECK AT THE WATER AND STAY HYDRATED WHILE KEEPING THEM DRY. ALSO, BY PUTTING IN MARBLES, YOU CAN PREVENT ACCIDENTAL DROWNING.

Base the size of your feeder and waterer on the number of chicks you have and brooder box size. Those too large for the brooder take away space from the chicks or make it difficult for them to move around. Those too small will need to be refilled frequently. Chicks need fresh feed and clean water at all times. Don't worry about exactly how much each chick eats and drinks, but rather provide a continual supply of both necessities.

Place the feeder and waterer apart from each other in the brooder box. As the chicks eat, they scratch through their food and toss it with their beaks. If the waterer is next to the feeder, some of this scattered food may end up in the water. Spoilage from this situation creates an unpleasant smell.

For young chicks, place the feeder and waterer directly on the brooder floor. As the chicks grow, raise the two to the level of their chests (but never any higher). Feed will always be spread about the brooder floor, but increasing the height of the feeder and waterer—by pulling hanging versions higher or adding small blocks or bricks to those on the ground—reduces the amount of feed they waste. Do not place the water directly beneath the heat source. Chicks need access to cool water, and the heat source will also heat the water.

Essentials: Bedding

Depending on the brooder, you may need to add bedding material to its bottom to help with cleaning. Its purpose is to reduce mess and (possibly) odor. There are several bedding options, some of which work better than others.

A good surface on which to stand will help strengthen chicks' legs and toes early in brooding. During the first day or two, fold up and place an old, nubby kitchen towel into the brooder box bottom. The nubby surface gives chicks a roughness to grab onto during the first few days

after hatching and prevents spraddle leg, a condition in which a chick's legs spread out from walking on a surface and on which it cannot gain traction. Spraddle leg is difficult to correct and can be painful for the animal. Once detected, immediately apply small splints, often made with toothpicks and plastic bandages. A chick with spraddle leg can die if you do not correct the problem in a timely manner because it cannot get to food and water. Most chick owners splint the legs themselves rather than going to a vet. For more about spraddle leg and other chick ailments, see chapter 10. Remove and wash the towel after it is soiled. After that, cover the brooder bottom with shavings in place of the towel.

Many people choose a wire floor for the brooder bottom. We recommend using ¼-inch (6 mm) hardware cloth instead. This allows you to place a pan or cardboard under the wire to catch any wasted food, spilled water, and feces. The only disadvantage to this is that it does not allow the chicks to effectively use their natural instinct of scratching for food.

There are several types of shavings from which to choose. Some studies have shown negative side effects of cedar shavings on mice and small rodents—though not on chicks—but as a precaution, we discourage their use. Straw can carry a fungus that your chicks can pick up. Try pine shavings, which are usually readily found and aren't expensive. Some still debate the safety of pine shavings, but we have used them many times and have never had a problem. Aspen shavings are said to be the safest; they release the least toxic gasses of any wood shaving, but are also more expensive.

Some people use shredded paper or newspaper as brooder bedding. Yes, newspaper is cheap, but it can harm the leg health of your chicks. When wet, these materials become slick and turn to mush, making cleaning difficult. Also, they can cause spraddle leg.

A rubber/plastic homemade brooder box with a wire lid

Everything (Else) You Should Know About Brooders

You may have additional questions before making a brooder decision. In spite of any advice given here or elsewhere, always evaluate all available options to make the decision that's best for you. Here are the answers to a few frequently asked questions:

HOW MANY CHICKS (FROM 0–6 WEEKS) WILL A BROODER HOLD?

While chicks start out egg-size, standard breeds quickly grow to softball-size around six to eight weeks old. What seems like plenty of room for a week-old chick will be quite tight for a six-week-old chick. Also, consider the space bulky items such as a feeder and waterer take up in the brooder. Though the heat source will not sit inside the brooder, it also affects the brooder's size and shape. Chicks need plenty of space to move toward or away from a heat source, at least 6 square inches (15 sq cm) per day-old chick, with 12 inches (30 cm) even more preferable.

HOW EASY IS CLEANING?

This may be one of the most important factors in your brooder decision. Cleaning takes time, and time is precious. Look for brooders easy to maintain and clean. You can easily rub down plastic and rubber surfaces with a cloth or rinse them out with a hose.

HOW FREQUENTLY WILL I NEED TO CLEAN THE BROODER?

The amount of space and number of birds in the brooder determine how frequently you need to clean it. Larger brooders require fewer cleanings than those with less space. Your nose and eyes easily let you know when it is time to clean out the brooder. We recommend doing it at least every other day. If you begin to smell ammonia, the level of gas is already too high for the birds' comfort. With the appropriate maintenance, you can keep odors low and reduce the chance for disease in your birds.

WILL THE CHICKS BE ABLE TO FLY OUT AFTER A FEW WEEKS? HOW CAN I KEEP THEM FROM GETTING OUT?

If you choose a brooder with low sides (such as the kiddie pool described on page 80), after the first week, chicks will flutter right over the edges. Just because they can get out doesn't mean they can always get back in— necessary to reach their food and water and for warmth. When they fly out, they also make messes where they aren't supposed to and may run into danger from household pets. If you brood chicks in the bathroom, you must keep down the toilet lid when the toilet's not in use to prevent drowning.

Even storage bins with tall sides may not prevent chicks from getting out. We have found that they flutter to the top of their feeder or waterer, which puts them at a height to then flutter out of the bin. Attaching upside-down cones to the tops of the feeders and waterers keeps the chicks from flying on top of them. Netting or a wire lid also may be a solution. Something as simple as a window screen will keep chicks inside the brooder and allow for good airflow.

Be sure to supervise young children when they approach
the brooder. Their curiosity could cause harm to your young brood.

DO I NEED TO PROTECT MY CHICKS FROM A HOUSEHOLD PET?

When you choose a brooder and its placement, keep in mind your pets—indoor cats, dogs, snakes, other household pets—and your young children. To protect the birds and keep unwanted intruders out of the brooder, clamp it down or lock it with a wire lid, shaped chicken wire, or netting. (Turn back to chapter 4 for more about this.)

WILL THIS BROODER PROVIDE EASY ACCESS TO MY CHICKS, FEEDERS, AND WATERERS?

The easier it is to access the items you need regularly in the brooder, the more enjoyable a brooding experience you will have. A brooder that requires you to bend down, stretch, or climb in will make the two months it takes for your chicks to feather seem much longer.

SHOULD I STORE MY BROODER FOR REUSE AFTER MY CHICKS OUTGROW IT?

If there is one confession chicken keepers make to each other, it's that chickens are addictive. Even if you think you won't add to your flock, you may change your mind. Think about where and how you can safely store the brooder and brooder items for future use; do not discard your brooder just because your first chicks grow up.

HOW EXPENSIVE IS A BROODER?

Brooders can be as cheap or expensive as you choose. Some people use a bathtub or a large cardboard box, at little to no cost. Others build a brooder using recycled materials, new materials, or a mixture of both. And still others purchase commercial brooders. There are options for everyone.

Placing Your Brooder

Once you decide on the materials for your brooder, you need to figure out where to put it. Chicks must stay warm, dry, and away from drafts. You can raise chicks in your bathroom or in your garage, but a sunroom, spare bedroom, or other interior space is ideal. These latter options likely include a heating unit and an air-conditioning unit which will help you keep the chicks comfortable no matter the season.

Ideally, you'll keep your brooder out of the elements. At the same time, a brooder filled with growing chicks for six to eight weeks poses its challenges. A friend once reported that though he kept his first flock of chickens in a brooder in the basement, the smell overtook the house—all the way up to the third floor. He soon figured out that the brooder sat in the same room as the return air vent for the heating system. Need we say more? Moving the brooder from one room to another solved his problem. That's just one scenario, of course, but don't underestimate how a dozen growing chicks might smell inside your home, especially in a brooder not properly maintained.

INTRODUCING YOUR CHICKS TO THE BROODER

The brooder is finally set up and your chicks are ready to move in. What's next?

(1) Remove the chicks from the transport box (either from bringing them home or moving them from the incubator) one at a time and gently place them in the brooder. As you transfer them, examine each chick for any deformities, lethargy, or feces around their vent area. (See sidebar, "Contending with Health Issues in New Chicks," right.)

(2) Once each healthy chick is inside the brooder, gently dip its beak in water. Allow it to throw its head back to swallow and then dip its beak in the feed.

(3) Repeat this process with every chick.

This is an important time with your chicks because not only are you showing them where to find their feed and water—what a mother hen would do—but you also are looking for any weakness that may cause problems for the chicks in the future.

WASH YOUR HANDS BEFORE AND AFTER HANDLING CHICKS TO PREVENT CONTAMINATION. YOU COULD INADVERTENTLY PASS SOMETHING TO YOUR CHICKS OR THEY COULD PASS SOMETHING ON TO YOU.

Contending with Health Issues in New Chicks

Pasty butt. Some chicks develop what is called "pasty butt," a collection of fecal matter around the chick's vent area. If this occurred in nature, a mother hen would clean the chick's vent. Now it's your job. Here's what to do:

① Fill a cup with warm water. Tear strips of paper towels into 3 x 4-inch (8 x 10 cm) sections.

② Put on gloves and begin picking up the chicks and looking at their vent areas. If you find a chick with pasty butt, dip a paper towel strip in the warm water and gently rub clean the area. The fecal matter often gets tangled in the chick's fuzz so you may need to carefully work on it to loosen it.

③ Check chicks daily for this issue because it can block the vent, constipating the chick, and potentially resulting in death. The problem usually goes away by the time the chicks are two to three weeks old, but it can last longer.

The good news is that not every chick will have this problem. Usually the chicks that need wiping today are the same ones that need it tomorrow.

Weak chicks. Any time we spot a weak or sick chick, we separate it immediately from the other chicks. We always have available a smaller plastic box that we call the ICU (Intensive Care Unit) brooder. The same food, water, bedding, and heat requirements exist for your ICU brooder as with your regular brooder. We separate these chicks for several reasons: Once healthy chicks spy a weak chick, they peck it, push it over, and run over it. In addition, this chick's problem may be contagious to the healthy chicks. Chicks that need to be separated include those that are lethargic or with runny eyes,

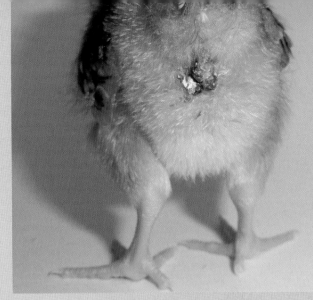

A chick with pasty butt that needs to be wiped and cleaned

diarrhea, or droopy wings. These chicks may not eat or drink. They are most likely quiet and want to sleep all the time. They may try to get as close to the heat source as they can, if they have enough energy to do so.

Set up the ICU brooder the same way as the regular brooder with food, water, and heat. However, it is usually smaller than the regular brooder and with an increased heat source (though one that doesn't overheat) from a bulb closer to the chick or a higher-wattage to increase heat. Chicks stay quarantined in this brooder until they are completely well or they succumb to their illness. Poultry vitamins and electrolytes found at your local feed-and-seed store may supply the chick an extra boost. In a pinch, if you can't get to the feed store, create an 8 percent sugar solution for the chicks (1½ cups [300 g] of sugar per gallon [4 L] of water) or put one dropper's worth of liquid baby vitamins (without iron) into the waterer base. Dip the chick's beak in the water tray occasionally throughout the day. It is a good sign if the chick drinks the water, but when it quits drinking completely, prepare for the worst.

When raising chicks at home, there is no mother hen to take care
of them, so as their owner, it's your job.

BROODER MAINTENANCE

Feeders and waterers always need a fresh and constant
supply. Depending on their size and the number of
chicks, you may need to refill them as frequently as
every two to three hours (especially homemade feeders
or waterers), daily, or every couple of days. Constant
monitoring of feed and water levels is essential.

It is also important to keep these essentials clean from
feces, which can contaminate a food and water supply.
To reduce this problem, keep them raised to the chicks'
chest level. Brood bantam and large fowl breeds
separately; the height requirements for food and water

equipment differ greatly. Attach an upside-down cone
to the top of the feeders and waterers to prohibit the
chicks from roosting and spoiling the feed and water
from above.

The brooder will need regular cleaning, the timing
of which will again depend on the brooder size, the
number of chicks you have, and the size and age of your
chicks. Here are some general guidelines: Spot clean
daily, (or at least every other day), and a full cleanout
should be done weekly. Clean as necessary in the
instance of a major water spill. Your nose and eyes will
also let you know when you need to clean.

Cleaning includes the following:

+ Replacing the bedding
+ Washing down the sides with soapy water
+ Taking apart and cleaning the feeders and waterers with soapy water and a sponge
+ Rinsing down the brooder with a hose

This may sound like common sense, but make sure to remove the chicks before you clean out the brooder. Place them in a box in a warm area.

When the chicks are ready to move outside to their new home, the chicken coop (which we cover completely in chapter 7), you need to clean the brooder one final time. We recommend ¼ cup (60 ml) bleach to 17 cups (4 L) of water. Use this bleach solution to scrub down the brooder, feeders, and waterers. Rinse these items thoroughly with water and let them dry. Store them safely for their next use.

Poster board is an efficient, disposable liner for your brooder.

ANDY'S ANECDOTE

I'M ALWAYS THINKING about how I can save time, especially when cleaning. Using poster board to line the bottom of the rubber/plastic brooder helps immensely. Cut a piece that fits the bottom dimensions. This allows quick bedding removal, which you can then easily dump into a compost bin. (See the photo of the poster board to go under the bedding at left.) Then I place a thick layer of my bedding choice on top of the poster board. When it is time to refresh the bedding, I grab the sides of the poster board and lift it out of the brooder, taking the soiled mess with it. A thick layer of bedding (about 3 to 4 inches [8 to 10 cm]) will prevent spraddle leg by keeping the chicks off of the slick poster board surface.

CHAPTER 7

HOME SWEET HOME: COOPS AND RUNS

NOW THAT YOUR CHICKS are ready to head outdoors, you need a coop, a structure that protects poultry from predators and weather and provides a dry, covered area to prevent feed and water from getting soiled or spoiled. You'll also need at least one run, a fencing structure outside the coop that protects chickens from predators and keeps them from wandering into inappropriate places—such as your neighbor's yard.

SELECTING A COOP

There are many types and styles of coops; which you choose for your backyard depends on your preference. Do you want it to look like a coop or a tool shed? Do you want one that complements your house? You'll have to look at this everyday, so make sure you choose one that's aesthetically pleasing to you. Like brooders, coops can be as cheap or as expensive as you choose. Following are a few options.

A model coop has a solid structure and a run big enough
to assure your flock security and exercise.

A movable coop has many advantages for your chickens, as well as the benefit to you in being able to control fertilization of specific parts of your yard.

This urban coop is beautifully appointed and illustrates the value of planning your structure inside and out before construction.

Chicken Tractors (Mobile Coops)

Chicken tractors, coops with either wheels or skids, are becoming popular in urban backyards. They come in all different sizes and styles, with a manageable starting cost, so there's likely one that's just right for you. Some have four wheels to allow for easy movement from one area in the yard to another. Some have two wheels in the back and handles in the front. This type of tractor can move around the yard the same way you move a wheelbarrow. Other tractors have skids (two large wooden boards) on the bottom; using a hitch, you can hook these up to a tractor or riding lawnmower to pull them from place to place.

Some chicken tractors are a single story with a run attached outside the screened area. Others are two stories with the run on the bottom.

There are many benefits to housing your chickens in a tractor comparable to those described above. It fits perfectly in a yard without much space or a large field. It effectively keeps your chickens safe and fertilizes your garden with the birds' poop. By moving the tractor one rotation every day, you can fertilize your yard, provide your chickens with fresh grass and bugs, and avoid creating the bare spot in your yard that a stationary coop does.

In fact, in our opinion, the only negative of a chicken tractor is that to avoid creating a bare spot where it sits, you have to keep moving it. (Andy has six bantam hens in his chicken tractor and has to move it every two to three days. It may not sound like much, but adding this step is sometimes too much, especially on rainy days.) The good news is that a chicken tractor can turn into a stationary coop if this works better for you.

Another benefit to a mobile coop is that it helps circumvent some city laws regarding poultry or building coops. Some require that permanent structures sit a specific number of feet or meters from a fence line or dwelling. But these laws may exclude a mobile coop by not considering it a permanent or accessory structure.

Stationary Coops

Stationary coops are just that: stationary. Some are small, which means you can move them as needed, but others are quite large. These coops don't have to be

A doghouse can be an ideal setup to get you started in keeping chickens.

expensive either. If you have recycling or cost savings on your mind, try cheap alternatives such as old storage buildings or greenhouses, a wooden barrel, or even a dog house made for larger breeds.

If you modify something you already had, think about sharp objects. Repair old wire, fix broken windows, and pick up any nails you see. Drag a large magnet along the ground to pick up nails, staples, screws, or any other small objects the chickens might accidentally eat. Repair holes and make the facility as draft-free as possible without losing adequate ventilation. Make special modifications for your specific breeds, for example long-tailed fowl or one of the many bantam breeds. Finally, run your hands over the surfaces to check for sharp objects.

Keep in mind that a clean coop is less likely to harbor disease or parasites. The smaller the coop, the more difficult it is to get inside and clean thoroughly—and the less likely cleaning will happen, period, especially if it means kneeling down in the mud or trying to move around in cramped conditions.

ANDY'S ANECDOTE

I USED DOGHOUSES in my run for many years before I ever had a built coop. My wife Jen and I still sometimes use doghouses for a variety of reasons, one being that we can move them around the yard easily. During the winter, we bring our chickens—and the doghouses—into the garden to let the animals fertilize the ground before the spring. When we use dog houses, I put hay inside across the entire bottom, and the hens lay their eggs in the back corners. No, you don't have to crawl inside to collect the eggs. Instead, use grabbers (tongs), or if your doghouse has a separate top and bottom, turn the top around to fit backwards on the bottom. With this option, you can reach in from the back and easily pick up your eggs. In the winter, the chickens huddle inside the dog house creating plenty of body heat to stay warm. In the summer, they roost on top of the doghouse. Many urbanites and suburbanites enjoy sporting coops with fancy designs or that coordinate with their houses.

PARTS OF A COOP

→ The most basic and essential parts of a coop rely upon the construction of a sturdy, protective, and weather-proof structure. If you are building your coop from scratch, be sure to build the coop large enough so the tallest member of your family can stand up in it. This makes cleaning the coop easier. Nobody likes to kneel down in dirty spaces that are too small to use a regular-size rake or shovel on cleaning days.

→ There should be one nesting box for each of your hens to lay her eggs in. Raising the nest boxes off the ground is a good plan. Attaching the nest boxes to the wall will free up some floor space.

→ Feeders should be placed in the coop or in a covered run to prevent rainwater from causing mold in the feed.

→ Keeping waterers outside in the run is fine, except during the coldest winter months when freezing may occur.

→ The best possible run you can build for your chickens is one where the outside space has a solid cover, as shown in the illustration, to prevent entry by wild birds and aerial predators.

→ A secure fence, one with a locking gate, should have wire at the base that is buried at least 18 to 24 inches (45.7 to 61 cm) underground to prevent entry by digging predators.

→ A coop, such as the one shown in the illustration, would typically have netting over the yard area to protect the flock from wild birds.

→ Windows are a great way to ventilate the coop, especially in the hot summer months. Be sure to cover the windows with chicken wire or ¼ inch (6 mm) hardware cloth to prevent the entry of predators. A good working latch is essential to prevent entry from nimble-fingered predators, such as raccoons.

→ Inside the coop, roosting poles provide a place for the chickens to rest at night.

Window secured with chicken wire and a locking system

Roosting poles

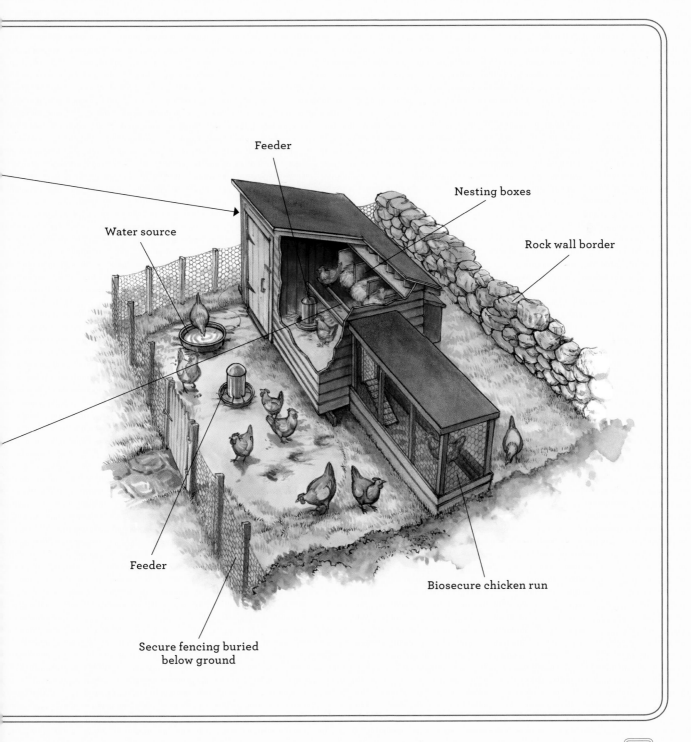

Feeder

Nesting boxes

Rock wall border

Water source

Feeder

Secure fencing buried
below ground

Biosecure chicken run

Best-case scenario, a coop should include six components: a nesting box, a roosting pole, a proper floor, a feeder, a waterer, and a run. As the example ideas for coops show, these requirements are somewhat flexible.

Nesting Box

A nesting box is the area where your hens lay their eggs. Most egg-laying chickens begin laying eggs between weeks twenty and twenty-four depending on the breed (some start as early as week seventeen; others take until week thirty), so ready your entire coop, including your nesting box, before then. Place nesting boxes inside the coop as low as ground level, but no higher than 3 feet (about 1 m) tall, and affixed to the coop for stability. It's also fine to stack them on top of each other. Though it is important to house the nesting boxes inside the coop, make them easily accessible to make collecting eggs easy and fun.

Nesting boxes can be constructed of metal or wood. You can construct your own or purchase them from many vendors.

ENCOURAGE YOUR CHICKENS TO LAY EGGS IN THEIR NESTING BOX BY PLACING GOLF BALLS, CERAMIC EGGS, WOODEN EGGS, OR ANYTHING THAT LOOKS EGG-LIKE IN A PLACE WHERE YOU WANT THEM TO LAY.

YES, CREATE AN INVITING EGG-LAYING ENVIRONMENT, BUT DON'T MAKE THIS AREA SO COMFORTABLE THAT THE HENS SLEEP THERE. PREVENT THIS BY ANGLING THE NESTING BOX FLOOR SLIGHTLY DOWN IN THE BACK. THIS WILL MAKE IT JUST A LITTLE UNCOMFORTABLE FOR THE HENS SO THEY'LL LAY THEIR EGGS AND MOVE ON.

Know that even if you provide your notion of the perfect nesting box for your chickens, when they first start to lay, they may choose someplace else. We frequently hear stories from frustrated owners who tell us their chickens are old enough to start laying but haven't yet produced eggs. We tell them to look around the yard; there *will* be eggs somewhere. We've heard of eggs found in old tires, flowerpots, and even pine straw piles. Once you locate them, pick them up and put them in the nesting box where you want your hens to lay. Hens will typically follow suit, laying wherever they see their eggs.

Most nesting boxes are built from wood in 12 x 12 x 12 inch (30 x 30 x 30 cm) squares, but there are many other options. For example, a milk crate, litter box, or other similarly sized plastic bin turned on its side can work well, or you can purchase metal nesting boxes.

You'll end up with a row of cubes with open fronts mounted to the wall, preferably with nothing stored beneath. If you are rearing large birds such as Langshans, Brahmas, or Jersey Giants, build your nest boxes larger.

The positioning of the roosting poles is key to for the comfort and safety of your flock. See the illustration on page 98–99 for more information about placement.

What might work for Lady Leghorn simply will not do for Miss Jersey Giant.

To protect the eggs from breaking once they're laid, the nesting box needs bedding material liner or a liner. A 2-inch (5 cm) layer of shavings works well. You also can purchase a nesting box pad from most online poultry supply stores. Avoid using straw. It can possibly cause health problems in chicks such as the fungal infection *Aspergillosis* (see chapter 10).

Roosting Pole

Every coop needs a roosting pole, a place where the chickens can sleep at night. This essential should be smooth and free from splinters to prevent injury that could lead to bumble foot. (We discuss bumble foot, a debilitating infection for chickens if left untreated, in chapter 10.)

Position the roosting pole 2 feet (60 cm) or higher off the coop floor, with soft bedding underneath. (Chickens jump off the roost each morning, so a cushioned layer can prevent injury.) An average roost measures 2 inches (5 cm) in diameter for standard breeds and smaller for bantams—a version of breeds approximately one-fifth the size of large fowl. On average, chickens need about 1 foot (30 cm) of roosting space per bird. If you need more than one roosting bar in your coop to provide all of your birds ample area, space the bars at least 2 feet (60 cm) apart. Position them at the same level or create a tiered effect.

Proper Flooring

The coop floor can consist of a permanent, wire, or dirt floor. Some floors are removable and can be pulled out of the coop, the contents dumped onto a compost pile, and the floor hosed off and returned with fresh bedding. Permanent floors are more difficult to clean because you have to scoop or shovel out old bedding and replace it with new material. Linoleum flooring (or a similar material) may be easier to clean than porous wood floors, but over time, may peel up. Wire floors also can work well. Either place the coop over your compost pile and allow the waste to drop directly in or set a tray under the floor that you can then lift out for easy clean up. If you go with a dirt floor, cover it with material such as woodchips for easier clean up. (Avoid hay and straw because they are nonabsorbent and can lead to high ammonia levels, which can damage birds' eyes and lungs over time.) Again, you need to scoop or shovel out dirty bedding for proper coop maintenance.

Both permanent and wire floors are quite secure when it comes to protecting your chickens from predators that dig. Know, however, that without a tray under a wire floor, some predators may try to eat or grab your chickens' toes and pull them through the holes. When designing your coop, be sure to protect the area from predators. (Chapter 8 goes into detail about predator protection.)

A secured area for ranging and socialization needs to be placed for maximum benefit. In addition to a stationary setup, you might consider a portable run.

Chicken Run

As discussed at this beginning of this chapter, the run is the fenced-in area outside the coop where chickens can range, scratch for bugs, and take dust baths. An optimal run offers sunlight and shade. The latter is especially important for chickens because they don't handle high temperatures well.

Consider the run carefully when designing the coop. Cover it with a solid roof or if that's not possible, netting such as 1-inch (2.5 cm), double-knotted aviary netting to keep out even the smallest of wild birds (and even many climbing predators). Wild birds carry diseases and parasites that can harm your flock. Also, keep the run away from birdfeeders to avoid attracting problematic wild birds that will want to eat your chickens' feed.

Also, bury the bottom edge of the wire surrounding the run at least 18 inches (45 cm) but as far down as 2 feet (60 cm) to prevent predators from digging in. If renting a trencher or digging holes in your yard is not an option, bending out the wire 18 to 24 inches (45 to 60 cm) from the edge has the same effect. Don't worry about aesthetics; grass will eventually grow over the bent-out wire, hiding it from sight.

Some runs have concrete flooring, but this isn't always the best option. Though it is easy to clean with a pressure washer, it eliminates the chickens' opportunity to forage, scratch, and take dust baths. Hay looks beautiful for the first few days or until it rains. Then it gets matted down, spoils, and produces an odor and moisture that can lead to respiratory distress for the birds.

Modern-day planning for coop-run design incorporates up to four separate runs where you can rotate the chickens. In one pen, plant winter wheat for them to enjoy in the spring. In another, plant sunflowers or corn for built-in summer shade. In a third, put in combinations of clover or other grasses that are hardy growers for your region. Rotating your hens in different runs that surround your coop ensures that they get the necessary exercise and enrichment they need without leaving behind a muddy, denuded pen. Many flock owners use this sustainable pen-rotation method. (For more on getting your yard ready, turn back to chapter 4.)

Regardless of whether you have multiple runs or just one, here's how to clean a run:

① Rake up chicken feces.

② Till the ground inside the chicken run to loosen the soil and create drainage. Add seed if doing a run rotation.

③ Add a layer of pulverized or powdered lime to the floor in excessively wet or smelly areas. The lime reduces any smell from the chicken waste. Block this off from the birds for a couple days. (Do not do this if you will be planting seed.)

④ Add a 3- to 4-inch (8 to 10 cm) layer of woodchips on top. This step is optional. We recommend large woodchips for the run because they do not biodegrade as quickly as the smaller ones do, meaning less frequent replacing required. Straw is not recommended for outdoor runs because it is nonabsorbent and will likely become matted.

ANDY'S ANECDOTE

WHEN I FENCE OFF areas of my yard for chickens, I initially leave the grass for them to eat. It won't take long for the chickens to flatten most of it to the ground. At that point, I cover the ground with woodchip bedding. (Call your local tree service, which will often dump a load of free woodchips right in your yard.) During the fall and spring, I put my chickens in the garden area or let them roam the yard to remove them from their runs.

It is wise to record the hours of sunlight and shade seasonally before making a final decision about the placement of your coop.

OTHER COOP CONSIDERATIONS

Once you have the main components of your coop in place, there are a few additional considerations.

Coop Placement

The placement of your coop is an important factor in creating a happy chicken home. By setting up the coop in a shaded area, you reduce the heat that accumulates inside on a hot summer day. Ideally, the coop should split half-sun, half-shade. If you don't have that choice, opt for full shade.

Here's another consideration: Do you have enough space for the breed to perform its natural behaviors without overcrowding? This becomes problematic with small coops, especially when it's too snowy outdoors for walking around. A coop is more than an overnight space in which to roost; it's where the birds remain when the weather outside is too dangerous or uncomfortable. For example, chickens dislike walking on cold surfaces. Therefore, when snow covers the ground outside, they may choose to stay indoors. Eventually they begin to feel "cooped up" and may peck at each other, pace, or stress out. These stresses open the door to disease. Even if snows last only a week, keep an eye on your birds for signs of coughing, sneezing, or stressed behavior such as feather pecking.

Also, chickens like to take dust baths. (Chicks as young as four weeks old will perform natural dust bathing behavior in the brooder if you give them a small container of dirt.) It is an innate comfort behavior for poultry. By dust bathing, chickens kick dust, dirt, or sand up against their skin, which helps them remove external parasites. You may wish to periodically provide chickens with a sand-filled container in which they can bathe. This will prevent hens from creating a dust bowl—a divot that can extend down as far as 20 inches (50 cm) in the bottom of your coop—that can cause you, the owner, many a twisted ankle. It is indeed not a pleasant situation.

Coop Temperature

To heat or not to heat? That is the question. We understand how cold it can get in some regions. Still, many people in colder climates don't heat or never have heated their coops. Each chicken produces a lot of body heat—as much as a 60-watt bulb—and these animals warm each other when they roost close together. As long as you provide chickens with a coop that protects them from weather elements such as wind, rain, and snow, chickens can cuddle and stay warm but may give up egg laying.

Different breeds have different temperature tolerance levels. (Turn back to chapter 2 for descriptions of several breeds.) Do your research before purchasing chicks to find a breed well suited to your area. The American Poultry Association lists large fowl breeds by region. If you live in colder climates, breeds developed in England, Australia, the United States, and Canada are cold-hardy. Some of these include Chantecler (the only Canadian breed), Australorp (the only Australian breed), Orpington, Dorking, Sussex, Plymouth Rock, Holland, Java, Lamona, Rhode Island White, Rhode Island Red, Buckeye, and New Hampshire.

Every large fowl breed of chicken has what's called a bantam counterpart (a smaller version). For example, Plymouth Rocks come in both large fowl and bantam. *Bantam* simply means "miniature" and does not indicate the name of a breed. Through the popularity of bantam breeds and the fact that you can raise more per square foot than large fowl, more breeds of bantams have been developed. Breeds such as the Dutch, Japanese, and Sebright only come in bantam sizes. These breeds are smaller and less able to keep themselves warm during the coldest parts of winter. Provide them with extra warmth for longer than you would for large fowl.

ANDY'S ANECDOTE

LIVING IN a relatively temperate region, I have never added a heat source to any of my coops. When people ask me whether to heat a coop, I always tell them the same series of answers: Chickens have been around since approximately 10,000 BC, but we have only had electricity for about 125 years and the chickens seem to have survived. Some reply with the following: "I know, but I will sleep better at night knowing my chickens are warm and toasty."

Remember, just about any heat source is a fire hazard so use caution when heating your coop. If you're worried about the cold temperatures, rather than heating your coop, a better choice might be to insulate it and protect your chickens from any drafts. Also, keep in mind the importance of ventilation and the fact that chickens will eat any insulation not covered by plywood or another solid building material.

While most people worry only about cold temperatures, hot temperatures also negatively affect chickens. That's why your coop needs good ventilation during summer *and* wintertime. Many people think it's best to seal up a coop in the winter. Some even jokingly mention wrapping the coop in a cling wrap. But having an airtight coop can cause more problems than cold for chickens. Ventilation helps control moisture buildup that when mixed with chicken waste can lead to mold growth and ammonia buildup. Good ventilation also reduces heat buildup and dust accumulation in the summer. Windows are effective vents, as are openings left at the top of the coop covered by hardware cloth. Incorporate at least two ventilation sections (one to let air in; one to let air out) to aid in airflow.

Feeders and Waterers

We discussed feeders and waterers in chapter 6, but in relation to brooders. You need these for the adult birds in your coop, too.

We encourage people to purchase the largest feeder and waterer available that they can afford and for which they have room. The best are those specifically designed for poultry, which your local feed-and-seed store should carry. Even with a small flock, the larger the feeder and waterer in your coop, the less frequently you need to refill them. Adult laying hens need 4 inches (10 cm) of feeder space and 1 inch (2.5 cm) of waterer space per bird, so factor in this detail when making these purchases.

When possible, place the chicken feed and water inside the coop because it protects the weather from soiling or spoiling them. Also, just as we suggested for the brooder, we recommend hanging the feeder and waterer in the coop (no higher than the chickens' back level). Chickens use their feet to scratch through feed, scattering it on the ground. They eat very little of this, which creates a lot of waste. Hanging their feed eliminates their ability to scratch through the feed and reduces the amount of waste. Leaving a waterer on the ground increases the chance the chickens may soil the water by stepping in it, defecating in it, or tossing woodchips up into the base. We typically place waterers on large patio blocks.

Extra Electricity

It takes about fourteen hours of daylight for a hen to produce an egg. On any given spring or summer day, chickens get plenty of sunlight. But as the days shorten and the nights lengthen during the fall and winter, egg production decreases. This, on top of the stress from cooler temperatures, may cause the birds to halt their egg production completely. Some poultry owners run electricity to their coops to give hens extra light to continue laying regularly. A household light bulb will do the trick. Use extreme caution with any electrical device in a coop, whether for lighting, heating, or cooling. All have the potential to start fires.

A supply of clean, fresh water is necessary for a chicken's healthy body systems and to maintain a consistent body temperature.

There are three schools of thought about when to run additional light:

First thing in the morning. If you set the light to come on early in the morning, you wake your chickens earlier, which may lead to more noise early in the morning. Often roosters crow with the first light. Even without a rooster, a hen reacts to light by eating, drinking, laying—and cackling. In urban areas, this may disturb you or your neighbors during sleeping hours.

Last thing at night. Turn on the light at sundown and continue lighting until your chickens have seen fourteen hours of light. Using this method, the noise the hens produce doesn't bother as many people and creates fewer problems all around.

Some in the morning, some at night. This may result in the same noise problem as adding light in the morning, just not as early.

Timers can be added to any light source so you can adjust when the light comes on or turns off based on your preference. They are usually inexpensive and easy to install.

Protection Against Predators

When designing and building a chicken coop and run, think about how to protect your chickens from predators. (For more about chicken predators, see chapter 8.) As we like to say, "There is always something that will love your chickens more than you." But you can take steps to keep your chickens safe. For starters, a solid floor made of concrete or plywood can help. A wire floor made from hardware cloth helps too, but some predators can pull the legs or toes of a hen through the bottom. The coop door should close and lock securely, and you should cover any windows or air vents with hardware cloth.

Runs are a little harder to secure against predators, but it can be done. Bury any fencing 18 to 24 inches (45 to 60 cm) deep to prevent digging predators from gaining access. For more protection, install poultry netting (somewhat expensive) or deer netting (cheaper but weaker than poultry netting) above the run/coop area to halt flying predators such as hawks and owls. A predator that tries to climb across deer netting will most likely rip it and fall right into the run. Snow can also cause problems with deer netting; as the snow accumulates and weighs down the netting, it can rip from where it is secured. You'll need to replace deer netting more frequently than poultry netting because it rots faster. Poultry or aviary netting is more permanent.

One of the best predator-proof coops we have ever seen was built when the chicken owner dug a 2 x 2 foot (60 x 60 cm) trench around his entire coop. He then placed a chain-link fence down into the trench, and poured in concrete. Next, he attached basic chicken wire to the inside of the chain-link fence to keep out smaller predators and pests. On the top of the run, he installed a chain-link fence to eliminate predators from the air, as well as climbing predators such as raccoons and opossums.

Make sure to consider your chicks when predator-proofing the coop. Chicks are little escape artists, plus predators such as rodents, snakes, and weasels take them without a second thought. Before putting your chicks into the coop, get down on your hands and knees to check for holes and then seal them up. Finally, count your chicks twice a day.

Never underestimate the power of a predator. Protection is the major consideration when constructing coops and runs. Install reliable locking systems and high-quality wire for security.

USE WIRE WITH SMALLER OPENINGS. NOT ONLY DOES IT KEEP OUT MICE AND RATS, BUT ALSO HAWKS, WHICH HAVE BEEN KNOWN TO WAIT FOR A CHICKEN TO WALK BY, SWOOP DOWN, AND REACH THROUGH THE FENCE OPENING FOR A QUICK KILL.

Inspect your coops and runs on a regular basis, and make repairs
in a timely manner to discourage predators.

COOP MAINTENANCE

Maintaining your coop does not have to be time consuming or difficult. As with a brooder, your eyes and nose tell you when it needs a good cleaning. Put your face down at chicken-head height for three to five minutes. If your eyes water due to high ammonia levels or you smell high ammonia levels, then you need to clean up. Without question, clean your coops inside and out twice a year, once at the beginning of spring and once at the start of fall. However, we recommend periodic maintenance between these cleanings, too.

To rid the coop of mold, bacteria, viruses, and other harmful organisms, here's how to do a thorough cleaning:

(1) Remove chickens and put them in a safe place.

(2) Take out the feeder and water to clean those pieces of equipment.

(3) Remove all bedding.

(4) Rid the coop of cobwebs and dust and then clean all surfaces with warm, soapy water. Let it dry overnight.

As far as inside the coop, check nesting boxes weekly for buildup of chicken waste. The nesting boxes should not contain much chicken waste; if they do, they're too comfortable and are acting as beds for your birds. If using plastic nesting box liners, hose off dirt or feces, scrub clean with a brush and soapy water, let dry, place into a bucket of disinfectant, and then let the liner dry before placing it back into the nesting box. Add fresh shavings if you are not using a liner, as needed, to protect the eggs. This also is a good time to look for mites (which we discuss further in chapter 10) or signs of them in the nesting boxes. Check the roost monthly for splinters or sharp edges that could injure your chickens' feet and lead to bumble foot. While cleaning inside the coop, look around the base of the trashcan that holds your feed for any holes created by mice or rats trying to gain access.

Finally, clean the feeder and waterer as needed. Use a hose and scrub brush to scrub out dirt, mold, or algae growth. Adding about ¼ cup (60 ml) of apple cider vinegar to a 5-gallon (19 L) plastic or ceramic waterer will reduce algae growth during the summer (although scientists have not yet fully analyzed the benefits versus the drawbacks of doing this).

Maintaining your coop properly prevents injury and disease in your birds.

Note: NEVER put apple cider vinegar in galvanized waterers—it will react with the metal and harm the chickens.

⑤ Clean it with a disinfectant the next day and then follow the label instructions, washing it off after the appropriate amount of time.

⑥ Rinse the coop clean, let it dry, and then put everything back in, including the chickens.

SAFE AND SOUND: PROTECTING YOUR FLOCK FROM PREDATORS

WHEN BUILDING YOUR COOP, think seriously about how to predator-proof the structure and any outdoor spaces in which your chickens roam. Perhaps skunks, raccoons, opossums, or foxes live in the area. Maybe you live near prime hawk nesting locations. It is your responsibility to prevent your chickens from living in fear that they'll be part of the evening buffet.

In many places, the law protects wild animals, putting your rights as flock owner second. For example, in the United States, it's illegal to shoot a bird of prey. Unless you are Native American, you cannot remove feathers from any dead wild bird. Why these protections? Because in the past, animal cruelty was a real concern. So although you may be angry that one of these animals attacked your flock, research your rights before you shoot or capture.

Your best bet is to contact your local animal control office or a government-affiliated wildlife biologist. With help from these officials, you can identify the culprit and determine an appropriate action plan. Though fees associated with local services may exist, they're likely minimal, but will ensure that the relocation of a captured animal follows regulations enforced where you live. Alternatively, if the predator has already made your coop an easy target, you may wish to contact a professional animal-removal service.

Much of the information provided in this chapter stems from our personal experiences with predators and chickens. But we also recommend the Internet Center for Wildlife Damage Management (http://icwdm.org/) as an additional resource. This website provides information about track identification, damage to poultry and livestock, and preventative measures to keep out predators.

⟦ Predator-proof every opening of your coop thoroughly with high-quality wire. ⟧

Secure heavy-gauge wire around your chicken run,
burying it 1½ to 2 feet (45 to 60 cm) below the frame.

PREDATOR-PROOFING

Later in the chapter, we get into specifics about different predators' calling cards and measures to keep each away from your coop. But first, here's some tips on making your coop a secure place for your birds. (Many of these we touched on in the previous chapter about coops, but they're important enough to reiterate here.) Remember to safeguard the coop from above, below, and on all sides.

When building a new coop, bury the wire on all outdoor runs at least 18 to 24 inches (45 to 60 cm) down. Most animals give up digging after awhile. As an alternative, bend the wire out into an L shape at least 18 to 24 inches (45 to 60 cm) from where it meets the ground. Does it look pretty initially? No, but eventually grass grows over it or gravel rocks you place around the coop base covers it.

If possible, cover all outdoor runs with solid roofing to prevent wild bird feces from dropping in and attacks from aerial predators such as hawks or owls. Also, a solid roof provides shade and shelter for your birds. If a solid roof is a no-go, cover the pen with 1-inch (2.5 cm), double-knotted aviary netting. (The 1-inch [2.5 cm] netting is better than the 2-inch [5 cm] because the latter allows wild birds entry into the shelter.) Place the netting as high as the tallest family member who enters the run.

When building your coop's indoor floor, whichever type you choose, install a ¼-inch (⅔ cm) hardware cloth beneath. This prevents entry by rodents that decide they

THE CHICKEN WHISPERER'S GUIDE TO KEEPING CHICKENS

want access to your coop by chewing through the floor. However, this preventative measure does not stop them from digging tunnels under the coop or making a home there. (You'll have to put out traps or bait should you find a rodent tunnel leading under the coop.)

PREDATOR CALLING CARDS

In many—though not all—cases of predation on a flock, the predator leaves behind telltale signs. As flock owner and therefore protector, it is up to you to examine the remains (while wearing gloves, of course) to determine which critter caused the damage to your flock. A predator who is successful the first time around will likely return.

The descriptions that follow may not seem characteristic of what you see on your farm post-attack. We've done our best to describe typical behaviors, but not every animal in a species acts in the exact same way. Also, young or sick animals out on their own may be more persistent or desperate in their feeding habits because of their extreme hunger, meaning they'll likely come back to a coop with minimal or poor predator proofing.

WOUND CARE

If your chickens are attacked by a predator, you need to act quickly. First, isolate the wounded bird—most likely the chicken will be in shock. Place clean gauze or a clean towel over any wound that is actively bleeding. Wrap the bird in a towel to restrict is movements and place it in a dark box (with air holes) so you can transport the chicken to your nearest veterinarian. Chickens have many delicate and thin-walled air sacs throughout their bodies that are a part of the bird's respiratory system. There is the risk of a predator puncturing one of these air sacs. Not only is there this risk, but birds are unique in their anatomy and medication requirements. Always seek a veterinarian's expertise in times of trauma.

Options to Protect the Flock

Do you need a little more protection for your birds? Try an automatic coop door. These operate on timers that slowly open and close the door at an hour of your choosing. Once the birds are in for the night, they're in, and you can rest easy knowing that your hens are locked up safely. The only problem with automatic coop doors tends to occur during the yearly daylight saving time clock changes. Most hens have never heard of daylight saving time and get caught outside the coop completely unaware that the door has closed early. During this time of the year, check that the hens all make it into the coop safely before dark.

Some flock owners swear that hanging shiny CDs, flashy Mylar strips, fake owls, or even fishing wire over their runs deters predators. Although these methods may work temporarily, predators are smart. Plus, they have all the time in the world to sit, wait, and watch. Eventually they may figure out your ruse and test the limits of your coop's boundaries.

Gray wolf

Coyote

Rottweiler
(domestic dog)

Canines

This predator group includes wolves, coyotes, and both wild and domestic dogs. Coyotes typically kill with a bite to the neck and disembowel the fowl. Wolves operate in much the same way. Dogs are less efficient and bite at the whole bird, but they do not consume much flesh. All three of these groups typically leave behind dead birds.

If you suspect a canine, look for footprints. A domestic dog's footprints vary depending on the breed but average about 4 inches (10 cm) in length. A coyote's are smaller (3 inches [8 cm]) and more compact. The footprints of a wolf measure approximately 5 inches (13 cm) long.

Domestic Cat

Red fox

Cats

Cats, both wild and domestic, are messy eaters that leave pieces of their meal strewn about in a wide radius from where they actually dine. They eat the meatiest parts of the bird, leaving the skin and feathers intact. They also tend to leave tooth marks on the birds' bones.

Foxes

Have you heard the phrase, "Fox in the henhouse"? It usually means that someone is up to no good—the precise behavior foxes, both red and grey, exhibit when going after your chickens. Foxes prefer poultry as a food source and kill with a bite to the neck. Usually the

animal leaves no evidence behind, taking the whole bird back to its den. All you may find are a few drops of blood or a few feathers.

Typically, a carcass left behind by a fox has curled toes because the fox stripped the meat from the leg bones. Or it partially buries the remains. A fox is a serious problem for your chickens because if it successfully raids a coop, the fox will surely come back the next day. Check for and repair any holes in the coop and look for fox tracks, which look similar to that of a dog, wolf, or coyote.

Opossum (or Possum)

Wild raccoon

Opossums

Opossums occasionally raid a chicken house, often taking a single bird at a time and clumsily mauling it. Opossums tend to feed on the bird starting at the vent or consume young poultry whole, leaving behind a few wet feathers. They'll also eat eggs left in the nest. You'll find eggs smashed and strewn about, often with just small pieces of shell remaining. Opossums may also come in just to eat the grain or feed in the feeder.

If you think you have had a recent visit by an opossum, smell around the coop. Does it have a musty odor? That's an indication of the animal. Also look for its tracks, which are unique, short in the front (2 inches [5 cm]) and long in the back (3 inches [8 cm]). The rear footprints have an elongated outer digit that looks almost like a thumb.

Raccoons

Raccoons are troublesome predators because they are smart problems solvers, returning night after night to test your coop's defenses, sometimes chewing through an outdoor run's aviary netting. (If for no other reason than to protect against raccoons, we strongly recommend a well-built coop and outdoor run with a solid roof covering.)

Raccoons enjoy a varied diet that includes meat, fruits, and vegetables. That means they'll certainly come after your chickens. These masked bandits have no qualms about taking several birds a night from your coop. How can you tell if a raccoon's been by? The breast and crop of the dead bird may be chewed on. Also the entrails may be eaten. Plus, raccoons are fastidious about cleanup, often using a nearby water source to wash up. In the

Weasel

Fisher

coop, this means the waterer, so look for bits of feather or flesh in or near your chicken waterer. Finally, raccoons also consume chicken eggs, taking them a short distance away from the nest to eat.

The tracks of a raccoon are distinctive. These animals have five toes arranged much like a human hand. Also, raccoons have an interesting gait, leaving tracks that show the left hind foot aligned with the right fore foot.

Weasels, Mink, Fishers, and Other Mustelids

In some places, mustelids (e.g., weasels, martens, minks, and fisher cats) are a problem for chickens. When they get into the coop, they often eat just a chicken's head, leaving the rest of the body intact. They can kill many birds in a single evening. And eggs are not off limits either. An egg eaten by an animal in this group typically has a ½- to ¾-inch (1 to 2 cm) opening at one end and possibly tooth marks. Look carefully at the egg to see the finely chewed ends.

Rats

Rats are inclined to take small chickens, especially chicks, killing the bird by biting at the back of the neck. They also tend to relocate chicks back to a burrow or hole to eat. A rat will consume less of a bird that's too large to fit in its hole, but will try to conceal the carcass somewhere nearby. Look for rat feces as an indicator that these animals are around.

Skunks

Skunks, as we all know, have a characteristic smell. If you enter a coop shortly after one has departed, you may catch a slight hint of its scent. Though these black-and-white animals, cousins of minks and weasels, tend to leave adult birds alone, they will take one or two smaller birds in an evening. Be aware of their telltale signs: They maul the birds a great deal. Or they take a bird's carcass back to a den site, leaving no trace. Skunks also eat chicken eggs, robbing a nest of its contents and moving no more than 3 feet (1 m) away to eat. An egg eaten by a skunk typically has one edge pushed in (this would have enabled the skunk to stick in its nose and lick out the contents). For incubated eggs, the skunk may chew them into small pieces.

Snakes

Snakes of all kinds do come into chicken coops, but they leave little evidence of their visit, often swallowing eggs or chicks whole. Sometimes they shed their skin, dropping it in the coop. They need an entry-exit point large enough to pass through after eating, so block all such entryways using rat wire or ¼-inch (⅔ cm) wire mesh. Finally, snakes may remain at the scene of the crime so reach into the nest box with caution. In other words, always look before you reach.

Rat scavenging

Skunk in early spring

Snake are excellent climbers and a particular threat to eggs and chickens.

Flying predators, such as this hawk, can swoop down in a matter of seconds. They are a major threat to your flock.

Aerial Predators

Hawks, eagles, owls, vultures, and other birds of prey can quickly and easily enter and exit a pen not covered with aviary netting. And if they don't perceive you as a threat—for example, if you're sitting reading a book and not moving much—they won't be afraid to swoop in and take birds only a few feet or meters from you in the yard. Eagles are large enough to carry a bird of any size; hawks and falcons can take all but the largest rooster and hen breeds. Vultures occasionally take chickens, though not too frequently. Owls hunt at night and fly silently. They feed mainly on rodents but can capture and carry off chicks with ease. If you notice your chickens being taken with great frequency at night, look for an owl nest nearby.

Some birds of prey, if they feel comfortable, will not fly away after capture. They will pull out feathers to reach the skin and flesh.

THE WALK-AROUND

Walk around your coop perimeter daily to look for predation attempts. Signs may be as obvious as a hole, scratches on the door, footprints, or feces. If you happen upon a predator in your coop, keep in mind that it is a wild animal. It may also be sick and the risk of it biting you is high if you've interrupted its mealtime. Remember, you may not be able to tell that an animal is sick simply by looking at it. The worst possible scenario is meeting a rabies-infected predator. No matter the scenario, allow the predator to leave quietly. If you come across a bird that has been attacked, use caution when handling the remains. Assume a sick animal was the perpetrator and always wear gloves to handle the remains.

Once a predator attack has occurred, focus your efforts on preventing future attacks and repairing the coop.

NUTRiTiON: COMMON MANAGEMENT QUESTiONS AND SUGGESTiONS

ALREADY IN THIS BOOK, we have discussed good coop design. Happy, healthy birds live in warm, secure coops. But, what about the second most expensive and most important part of keeping chickens, the food? Feeding a hen scraps, scratch, and leftovers will not help her to live a long and healthy life or reach her potential in egg production. With high-quality feed and a consistent diet, chickens can live to be ten years old. Some even make it to fifteen.

The human race knows more about the dietary requirements of the modern chicken than we do about any other livestock animal. Not that it's been easy to learn. Feed formulations are complex and the specific needs of chickens change during different life stages. Feed mills even hire poultry nutritionists to formulate their food rations to get it just right.

Thousands of researchers, through myriad experiments, have clearly identified the dietary needs of commercial chickens. Your chickens at home may not be the same breeds as those used in commercial production and therefore likely do not need to produce eggs at optimum performance levels. Regardless, you still should understand the basic requirements of a chicken's diet. Chickens are amazing creatures; with the right building blocks, they can produce for you an egg a day during their first year or so. But that requires the correct ration formulation to allow the chicken to convert feed appropriately for its body.

The primary components of a chicken's diet are as follows:

- ❖ Water
- ❖ Protein
- ❖ Carbohydrates
- ❖ Fat
- ❖ Minerals and vitamins

[Free-range fowl: a healthy specimen]

Without a constant supply of clean water, a chicken will stop laying eggs.

WATER

Water is perhaps the most overlooked yet most essential nutrient in a balanced diet for chickens. On average, every day, these birds consume twice as much water as they do feed. Clean water free of algae and bacteria is essential to the overall health of their gastrointestinal tract and the associated immune system.

Chickens use water to lubricate their body systems and for temperature control. The liquid makes up components of their muscle, blood, and bone, plus 66 percent of the eggs we eat. A laying hen, by percentage of body weight, is 62.4 percent water—higher than the percent of water found in most four-legged livestock animals. Without adequate fresh, clean water, the hen has no choice but to stop laying eggs.

Not just any water will do. Keep in mind that the average chickens cannot access water that's frozen over. (They won't peck through thick ice to get to water below.) They also avoid water that's too hot. A good rule of thumb: Water too hot for you to drink is too hot for the chickens, too. During winter, keep your water warm and during summer, keep your water cool.

PROTEIN

The building blocks of protein are amino acids. About half the protein in the average poultry diet comes from soybean meal or other related grains. (Your feedbag label should list soybean meal as one of the two primary ingredients.) The other half comes from high-protein supplements such as plant- or animal-sourced byproduct meal. (It's exactly what it sounds like.) What's most

Soybeans are a major source of protein
in your chickens' diet.

A balanced diet will ensure that your chickens will
receive all nutrients in the correct proportions.

important when it comes to protein in the poultry ration is the balance of amino acids.

A diverse diet will help ensure overall health.

Supplementing the diet with proteins from multiple sources simulates a chicken's diet in the wild. Remember, chickens are by no means vegetarians.

Plant proteins include the following:

- ⤳ Oilseed meal
- ⤳ Soybean meal
- ⤳ Coconut meal
- ⤳ Cottonseed meal
- ⤳ Linseed meal
- ⤳ Peanut meal
- ⤳ Rapeseed/canola meal
- ⤳ Safflower meal
- ⤳ Sesame meal
- ⤳ Sunflower seed meal

Animal proteins the following:

- ⤳ Meat-packing byproducts (meat and bone meal, blood meal)
- ⤳ Feather meal and poultry byproduct meal
- ⤳ Dairy (dried whey, dried milk protein, dried skimmed milk, dried buttermilk, and dried milk albumin)

- ⤳ Marine byproducts (fish meal, fish liver, and glandular meal, shrimp meal, crab meal, dried fish solubles, and whale meal)

A poultry nutritionist at the mill that provides your feed analyzes each component for its amino acid content and adds in appropriate amounts to create a perfect balance for your chickens' diet. As a chicken owner, this is important for you to understand. However, nutritionists are hired to ensure that balance, so it's the rare occasion that requires you to know much more about this topic.

SOYBEANS CAN WORK WELL AS PART OF YOUR CHICKENS' DIET, BUT KEEP IN MIND THAT SOYBEANS CONTAIN AN IRRITANT CALLED TRYPSIN INHIBITOR. CHICKENS CANNOT DIGEST THIS, AND IN FACT, IT PREVENTS THEM FROM ACCESSING THE SOYBEAN'S ESSENTIAL NUTRIENTS (GOOD DESIGN BY A PLANT LOOKING FOR A SEED TRANSPORTER). TO INACTIVATE THE TRYPSIN INHIBITOR, ROAST THE SOYBEANS FIRST.

Wheat is one of the major ingredients of a balanced diet, especially popular in Canada.

Provide a well-balanced diet including vitamins and minerals.

CARBOHYDRATES

The word *carbohydrate* is synonymous with energy when it comes to poultry-diet formulation. Chickens eat to meet their energy needs. Simple sugars, also called monosaccharides, make up the building blocks of carbohydrates. Glucose, fructose, mannose, and galactose are monosaccharides, and one of the only types of energy chickens have the enzymes to break down. They can't break down cellulose or lactose. But don't many people let their chickens eat grass or yogurt? Yes, indeed, they do. Chickens do not have a ruminant stomach like cows do, but they may gain many of the benefits of eating grass. Nonetheless, they like it. Yogurt, is a cultured, dairy-based product and a good source of animal protein and probiotics. Chickens cannot digest lactose and you should avoid feeding them uncultured dairy products, such as milk, which come with a major side effect: diarrhea.

In the United States, corn is the typical energy source for chickens. But other grains may also be fed to your backyard flock. In Canada, for example, chickens tend get their energy from wheat. Be mindful of the grain source you choose. Wheat can sometimes be prickly and irritate the chicken's digestive tract, which can lead to gut-lining damage and a host of other problems.

FAT

In feeds, fats function similarly to carbohydrates, providing a source of heat and energy. When used as fuel to produce heat, fat gives off more than two times more heat than carbohydrates, so the diet requires far less fat to accomplish the same outcome.

Poultry will not eat certain low-in-fat feeds because they lack the appropriate taste and texture. Also, fat acts as a vehicle to evenly distribute certain micronutrients. Fats contain fatty acids, two of which are essential: linoleic acid and arachidonic acid. Without these, birds may grow poorly, have fatty livers, lay smaller eggs, and have poor hatchability.

If you choose to mix your own feed, keep in mind that fats can go rancid and need to be stored carefully. This should factor in to the calculation of feed mixtures. However, again, we recommend you purchase already-mixed feed.

MINERALS

A diet deficient in the appropriate minerals and vitamins can, for chickens, lead to drops in egg production, leg problems, or even weight loss followed by death.

Minerals are key to chickens' skeletal formation, hormone building, enzyme activation, and osmotic balance (balance of water to solids in the cells of the body). The following three minerals must make it into their diet.

Salt

Chickens have 24 taste buds, (compared to 350 in parrots, 9,000 in humans, and 15,000 in pigs), and salt is one of the few flavors they can indeed taste. It's also important for bone development, eggshell quality, and growth. But be careful of incorporating too much into their diet; high salt levels will give the birds diarrhea. They'll attempt to regain osmotic balance by drinking lots of water. Salt should make up between 0.2 and 0.5 percent of the diet.

Calcium

Every good hen keeper knows that adequate calcium in a diet equates to strong shells and solid bone formation. A hen needs, at minimum, 3 percent calcium in her diet and usually that comes from the balanced laying-hen feed. Provide a hen that requires additional calcium a supplement of oyster shell, which you can purchase by the bag or in bulk. Place this in a feeder in the coop. Chickens that need it will feed from it.

Vitamin Discovery

It was in the chicken that the first vitamins were discovered. On the island of Java, Christiaan Eijkman discovered vitamin B1, also known as thiamin. Henrik Dam discovered vitamin K by using chickens. For many years, the scientific community measured vitamins in "ICUs" or "International Chick Units" and indicated as such on the label.

Phosphorous

Phosphorous is essential to birds because it helps break down and use carbohydrates and fats they get from the feed. Phosphorous, in a balanced ration, often comes in the form of dicalcium phosphate or even steamed bone meal.

VITAMINS

Vitamins fall into two categories: water-soluble and fat-soluble.

Water-Soluble Vitamins

Water-soluble vitamins include biotin, choline, folacin (or folic acid), niacin, and the B vitamins (B1, B2, B3, B6, and B12). Normal activity and body action remove water-soluble vitamins from a chicken's body, so these need regular replacement through diet or supplements placed in water.

Fat-Soluble Vitamins

Chickens store fat-soluble vitamins in their body fat, so they need replenishing less frequently than water-soluble vitamins. The most important fat-soluble vitamins help the chickens build the most valuable fat-soluble vitamins: A, D, E, and K or A-D-E-K ("build a deck" is a common mnemonic device). These vitamins are essential building blocks to normal life. Appropriate amounts of all of these vitamins come in pre-mixes, sold in the form of balanced diets.

VITAMIN A

Vitamin A deficiencies affect chicks more than adults because adults have larger fat stores. Also be careful of vitamin A toxicity. However, this occurs only when chickens receive 500 times more than they need.

VITAMIN D

Vitamin D is often listed on the chicken feed tag as *cholecalciferol*. Only in the presence of sunlight does cholecalciferol convert to something useable by a chicken's body. Hence, vitamin D is often dubbed the "sunshine vitamin." Without adequate vitamin D, the chicken can suffer from rubbery bones or beaks otherwise known as rickets.

VITAMIN E

Vitamin E is essential to build strong muscle and nerve tissue, proper operation of the circulatory system, and good hatchability (important if you breed for chicks).

VITAMIN K

Vitamin K is named as such because it's essential to coagulation (*koagulation* in Danish, the language of Denmark, the country in which the vitamin was discovered). Chickens require vitamin K even more than usual when they suffer from internal parasites such as roundworms or *Coccidiosis*. If starting chicks without medicated feed (coccidiostats) or if a flock suffers from *Coccidiosis*, the risk of intestinal hemorrhage or increased clotting time increases due to damage caused by the burrowing action of the parasites.

OTHER FEED CONSIDERATIONS

As flock owner, if you mix your own feed, your biggest feed-related concern may crop up when a pre-mix goes bad or expires. When buying a pre-mix or supplement, look at the expiration date and choose the freshest mix available. By now, you may think that creating a poultry ration is complicated. It may sound like rocket science, but it's actually just poultry science.

There are also feed expenses to consider. In the United States, corn in chickens' diet provides energy while soybean meal provides protein. As corn has been diverted to ethanol production and its prices have skyrocketed, chicken feed prices have increased correspondingly. Many small flock owners are lured in by the low cost of chicks and ease of ownership, but can feel overwhelmed by the price of feed they must purchase every month or two. Factor in the feed cost when considering adding a flock to your family.

As we mentioned, it's crucial to feed your chickens a specific, balanced diet. We have seen many first time flock owners buy the most inexpensive feed—usually made of scratch grains or scratch that come from a mix of cracked corn and other seeds—for their chicks. This is *not* a balanced diet and should be considered candy or a treat for your flock. Another form of candy for your chickens is table scraps. Remember, give all treats in moderation. If your birds learn to expect them daily, they will stop eating their balanced diet in favor of waiting for what you hold in your hand. It's reminiscent of a five-year-old who wants ice cream after dinner, but who refuses to eat his broccoli.

The price of corn these days is not "chicken feed," causing prices to increase dramatically.

A high-protein starter diet, either medicated or nonmedicated, makes for a good beginning for your young brood.

DIETS FOR ALL AGES

Chicks have different nutritional requirements than adult chickens. The younger ones require more protein to build muscle, bones, and many other body tissues. Their diets, usually called chick starters, come in several different forms, sold as either a mash (a powdery mixture of chicken feed) or crumble (the same mixture in pellet form and broken down into pieces small enough for a chick to eat). These diets should include about 20 to 22 percent protein—a level maintained for six weeks.

You'll also have the choice to buy medicated or nonmedicated chick starter. Read through chapter 10 before deciding to make sure you understand the health risks associated with giving your chicks nonmedicated feed. For example, *Coccidiosis* is an internal parasite chickens get that can often be prevented with medicated feed. In general, medicated feed is a good choice for chicks raised in the house in a brooder, because they won't be exposed to native cocci and won't naturally develop immunity. Chicks raised on the ground, whether with

their mother or in a brooder with access to the soil, have the opportunity to be exposed early to native cocci and will develop immunity naturally. Remember, making an informed decision makes you the best possible flock owner.

Between six and sixteen weeks, a growing hen needs a grower diet that includes lower levels of calcium than that for laying hens and protein levels lower than in a chick starter. (This is also the diet a rooster will eat for most of his life.) The grower diet should provide 16 to 18 percent protein for these birds and about 1 percent calcium.

For a mixed flock (roosters, laying hens, different ages of birds, etc.), you can feed them all grower feed and offer the hens oyster shells for calcium. (This is after the chicks no longer need chick starter, of course.) It is very difficult to ensure each member of the flock is eating his or her own food!

After sixteen weeks, a laying hen should eat a diet with 16 to 18 percent protein, but it must also contain the appropriate levels of calcium for eggshell formation, with a minimum level of 3 percent calcium. Feeding the wrong diet may cause unintentional harm to your birds so make sure you give your flock the appropriate feed for the animals' age group.

SCRATCH FACTOR

Sometimes it's fine to use scratch. However, many chickens prefer its taste to that of normal feed, so use it appropriately or risk your hens' overall health and welfare.

One fine use of scratch is to give your birds an energy boost. In the winter months, when temperatures at night drop, throw a handful or two to your birds—about one-third to one-half of a handful for each adult hen—right before they roost. They'll use the extra energy overnight to maintain warmth. You'll know you've given them too much if they don't clean it all up within fifteen to twenty minutes. Touch your hens and feel for excess fat every other week. If they're gaining too much weight, cut back on the amount of scratch you give.

Birds on a low protein, high-energy diet, which can result from too much scratch, can become aggressive toward one another and show signs of obesity and feather-eating.

Birds on a low protein, high-energy diet, which can result from too much scratch, will eventually exhibit signs of obesity, feather-pecking, and feather-eating (which, in turn, may lead to cannibalism as the birds go after each other for their feathers). A bird's skin should feel loose and flexible, not tight as a drum. Overweight birds may not perch on high roosts given that their balance is compromised. Instead, they'll sit on the floor.

They also may begin to peck and eat feathers off of themselves or each other, to get the additional protein their bodies are craving. You'll know this is happening if you never see extra feathers floating around the pen. Of course, loss of feathers during an annual molt is normal and expected, but pay attention if it's the wrong time of year for molt—chickens typically molt during summer, but every bird is different—or if you notice more feathers than usual. Birds eat feathers for the keratin protein they contain, the same as found in human hair and fingernails.

This is *not* what was meant by, "a chicken in every pot ..."! Make sure all of your chickens are eating their fair share. Check their crops at the end of the day to make certain they are full and satisfied.

Breaking their bad habits early helps in the long run (although you may endure unhappy looks for a week or two). Here's what you do: Once you identify the problem—table scraps, junk food, scratch, and so on—remove it from the diet. Keep your chickens on a diet balanced to meet their specific needs (for example, layer pellets for hens in lay). These diets, although produced commercially, have been perfectly balanced to meet the chickens' dietary requirements. Expect your chickens to act like petulant children, not eating their normal food, holding out for the regular treat. This will test your endurance and patience, but your hens will eventually eat the balanced diet and slowly, but surely, they will stop holding out. Follow this dieting formula for six months.

After a half-year, reassess your hens' overall health by weighing them and feeling their overall body fat levels. Nobody knows your girls like you do! Once their weight loss plateaus, consider reintroducing scratch in small amounts, perhaps once a month. The same goes for table scraps. Remember, if obesity and feather-eating occur in your flock, even in the slightest, it's your job to correct the situation. Give all treats in moderation.

COMMON NUTRITION QUESTIONS

Even after you know what to feed your chicks, you may run into difficulties. Here are answers to some questions that may come up:

Are my birds eating their new feed? If you wonder whether your flock members are eating enough, here's a quick and easy method to find out: Birds that are well fed fill their crops (the outpocketing for the esophagus in the neck region just above the breast) during the last meal of the day, just before roosting. Once your flock roosts for the night, go in with a flashlight and lightly touch each hen's crop. You'll be able to feel feed in there. Do not push too hard as this can be uncomfortable for the bird.

The guy at the feed store is trying to sell me vitamin C. Should I buy it? Chickens, like many animals, can produce their own vitamin C. In fact, humans, some primates, guinea pigs, and spiders are the only animals on our planet incapable of making their own. So if the feed store salesman tries to sell you a supplement high in this vitamin, move on. Chickens don't need it; they make their own!

My hens seem to be losing breast feathers? The exposed skin is thick and hot. Are they okay? Many chicken owners believe this condition has to do with nutrition, but it actually deals with incubation. In the spring and summer, hens begin thinking about sitting on eggs. The changes to their bodies indicate development of a brood patch, or an area on the breast where the eggs come in direct contact with the hen's skin for incubation. This is normal and soon the hens will want to do nothing other than sit in the nest and incubate eggs. A hen that goes broody—the overwhelming, hormone-driven desire and willingness to sit on eggs—does not actually lay eggs, so you may wish to break her of this habit to get her laying again.

Some hens sit on eggs far past the twenty one–day incubation period. Watch these birds closely to ensure they eat a balanced diet and drink enough water. If broodiness continues far past twenty-one days, then break the hen of the habit. On a warm day, fill a sink with ice-cold water and dunk the hen up to her neck for up to a minute. Let her run free in the yard to dry off. Keep her away from the nest boxes for the next several days. You may need to repeat the process a couple times. Your hen will not be happy with you, but at least you will get her body back into a normal hen routine.

I want my chickens to live a natural life, so I want them to eat a natural diet. Does that mean I should give them organic feed? That is a decision that you need to make on your own. There is not yet an official definition for *natural* when it comes to poultry management, much less feed. The only way to guarantee truly organic feed is to purchase products with the "certified organic" label. To legally be called as such in the United States, for example, those feeds have had to undergo rigorous evaluation and specific processing to determine whether they meet specific qualifications. Certified-organic feeds cost more because growing grains this way is more expensive.

If you allow your birds access to pasture, be aware that also means access to plants and bugs, which is their natural diet but may not be "certified organic". (Chickens do not make the enzyme necessary to fully break down and utilize cellulose. Because of this, chickens will not fully benefit from pasture in the same way a cow does.)

CHAPTER

10

iLLNESS AND AiLMENTS: PREVENTiON AND TREATMENT

NO FLOCK OWNER WISHES illness on any flock member. Chickens are like any pet or member of your livestock family, meaning you should always have their health in mind. When the family cat or dog becomes ill, what do you do? Head to the veterinarian, of course! It is an expectation and responsibility that comes with owning a pet.

Because chicks cost so little to purchase and maintain, some non-chicken owners may consider them expendable. But you, on the other hand, likely consider them integral members of the family. Know this then: Veterinary care for poultry—or any avian species, for that matter—is among the most costly, and poultry diseases are some of the most difficult to diagnose. Therefore, preventing your chickens from getting disease should be a top priority.

Chickens, like all of your livestock, need reliable medical care.

BIOSECURITY

Prevention comes in the form of a somewhat-intimidating word: biosecurity. This term simply means "life" (*bio*) and "safety" (*security*). For flock owners, it means a daily set of preventive practices done to reduce the risk of disease transmission to your flock.

Regardless of your flock size, do not let this word or the measures it requires alienate you. In fact, owners with smaller flocks are potentially more powerful in enforcing biosecurity tenets than many large-scale livestock producers. Simply put, you have fewer birds, you know their favorite activities, and you interact with them daily. Commercial operations still use the same principles, but apply them on a significantly larger scale. You can control exactly how much exposure to disease—if any at all—you are willing to take.

Disease-Carrying Entities

How can you ensure that your flock members remain free from illness? How does disease enter your flock? Once you know your disease-causing enemies, are you willing to shut down all access points? To start, let's look at some of these disease-carriers.

WILD BIRDS

Though they are not predators of your flock, wild birds are some of a healthy flock's greatest enemies. As flock owner, it is your responsibility to prohibit contact with wild birds as much as possible. This is when many owners become disenchanted with the idyllic notion of keeping chickens. Doesn't every owner dream of letting her chickens roam free? Aren't the birds happier and healthier when they do? True, chickens enjoy adventure, but remember, they are birds that originated in the bamboo forest in Southeast Asia so they crave a certain amount of protection.

Turkeys, ducks, geese, and chickens: Keeping multiple species together is an in invitation to disease that can cross the boundaries of a single species.

Wild birds' feces not only carry countless viruses, bacteria, and protozoa, but their bodies also host internal and external parasites.

Here's why that's problematic: Many wild birds are opportunists, which means they'll take advantage of an appealing-looking coop for both shelter and food. Your biosecurity goal is to prevent these behaviors. In other words, stop wild birds from roosting in the coop, stealing feed meant for your chickens, and potentially leaving behind disease-laden feces.

To do so, fence out the wild birds using chicken- or pigeon-wire and solid roofing. Relocate birdhouses and feeders so they're as far away from your coop as possible.

Consider birdfeeders permanently disease-prone, to be handled *after* daily care of your chickens. If wild birds do get into your coop, remove their feces as quickly as possible; chickens naturally peck at interesting objects—that can include wild bird feces.

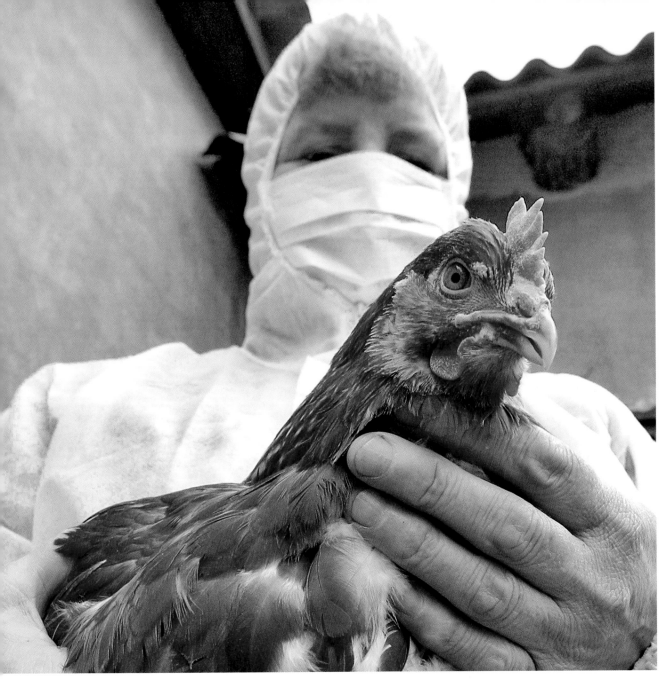

Either you or your veterinarian can vaccinate poultry.

RODENTS

Once rodents find a safe place with plenty of food, they stay put. This is problematic because they carry organisms in their hair and feet and their intestines produce thousands of disease-causing organisms. (Interestingly, a mouse does not have an anal sphincter, and therefore leaves droppings around its most highly traveled routes.) Mouse feces contaminate feed and water sources, and chickens eat these feces out of curiosity. In a single twenty four–hour period, one mouse can excrete, on average, 100 fecal pellets, each of which may carry up to 230,000 *Salmonella* Enteritidis organisms (we get into more detail about salmonella on page 147).

Keep food inaccessible to rodents. That means cleaning up spills and storing feed in a metal trashcan with a tight-fitting lid. These bins keep out rodents for a time but not forever.

Rodents have incisor teeth that never stop growing, requiring that the animals grind them down by chewing. They can chew through wood, cinder block, and yes, metal, so check the feed room, the coop, and your metal trashcans weekly for holes. Mend rodent holes immediately, with ¼-inch (⅔ cm) wire mesh.

Also, examine your coop and feed room for rodent feces and remove them immediately; then wash your hands.

PREDATORS AND DISEASE

Predators not only bring in zoonotic organisms on their feet and fur, but also in any feces they leave behind in the coop. (In chapter 8, we discussed predator-proofing your coop.)

The biggest concern for the average flock owner is a predator found within the coop. Although chickens cannot get rabies, a human can. If you come across a predator in your coop, leave it alone. Never attempt to remove it or you risk being bitten.

Once the predator leaves, you need to remove any bird carcasses. Always wear gloves to do so. If a chicken that was attacked survived the ordeal, remove that bird to your quarantine area until it is again well. Always err on the side of caution.

Disease organisms from predators can survive in the earth and become a serious threat to your hens and chicks.

HUMAN SHOES AND CLOTHES

What evil lurks on the *soles* of men? Dust, dirt, rocks, mud, even leafy debris can surround and even protect a virus, bacterium, fungus, or other disease-causing organism. The same goes for clothing. You and your guests should change out of street clothes before entering your coop. Everyone should walk through a disinfecting footbath to prevent a disease from waltzing right in.

Keeping Disease Out

Now you know some of the ways diseases can get into your coop. Let's discuss how to keep them out.

PHYSICAL BARRIERS TO ENTRY

It is your responsibility to avoid exposing them to disease by providing them with great living conditions. A well-built coop with a secure door protects against diseases brought in by wild birds, rodents, predators, and uninvited humans. Part of this includes a fence, which not only keeps your flock in check, but also keeps out unwanted predators. A good coop protects your flock from ground and airborne predators. (For more information about specific predators, turn back to chapter 8.)

COOP-SPECIFIC CLOTHING

As we mentioned above, disease-carrying agents can cling to clothes and shoes. Have dedicated clothing for when you feed and care for your flock. Wash these items weekly. Also, if possible, use a dedicated pair of boots that you can scrub weekly or wear in a footbath. A footbath can be made from a container such as a litter box filled with a disinfectant solution, and you can walk through it to prevent tracking in disease. You can always wear coveralls when you visit the coop.

A DISEASE-CAUSING ORGANISM THAT CAN MOVE FROM ONE ANIMAL SPECIES TO ANOTHER OR TO HUMANS IS CALLED A ZOONOTIC ORGANISM. THAT MEANS IT IS CAPABLE OF INFECTING AND GROWING IN MORE THAN ONE TYPE OF ANIMAL.

A well-built coop is worth the time, planning, and cost to ultimately protect your brood. Secure doors and windows, a functional locking system, proper lighting, and ventilation are only a few of the amenities necessary to consider in your overall design. See the illustration on page 98–99 for more detailed information.

QUARANTINE AREAS

Quarantine all new birds for two to four weeks in isolation before placing them with your flock. This ensures that if a new bird is carrying or comes down with a disease, you will not immediately expose the rest of your flock to the problem.

It is never a good idea to introduce new birds to an already existing flock because some birds may be disease-carriers without exhibiting symptoms. Those birds can still infect your flock. The more closed you keep your flock, the less chance you have of letting in disease.

CLEANING

Disease does not need an animal host to gain entry to your flock. Dirty or contaminated equipment can house disease organisms. Clean and disinfect all equipment that enters your property. For that reason, we advise against borrowing a coop or carrying cage without thoroughly cleaning and disinfecting it first. Here's the bottom line: Keep clean everything that touches your birds.

TRAFFIC PATTERNS

If you have experienced disease outbreaks in the past, ask yourself, "Did I do something to cause this?" Think about it like this: Usually, everyone's first morning task is to grab the food and feed the birds. That means everyone heads first for the feed room (which makes it a great place for a footbath, by the way) but where to next? Perhaps after leaving the feed room, rethink your normal path instead of caring for the nearest birds.

Feed and take care of your youngest birds first, and then move to the older birds. Though it may go against intuition, care last for the birds held in your sick pens or in quarantine. This prevents spreading the disease further among your flock members. Keep the equipment for birds in the quarantine area completely separate and make it obvious to others who handle the chickens (for example, place a red piece of tape or a red marker around quarantine area equipment).

DISEASE DIAGNOSIS

Birds hide illness as long as possible. It's in their nature. In the wild, predators dispose of birds that show signs of sickness. If not killed by predation, these birds are ostracized by their flocks. They also are the last selected for mating purposes. Therefore it behooves them to look and act as normal as possible for as long as possible.

By the time you notice illness in a bird, it may have already disseminated virus, bacteria, and other germs to the rest of the flock. This bird still needs you as its advocate.

First, isolate it immediately.

Examine the rest of your flock for signs of illness.

Finally, work toward diagnosis and then treatment for the sick bird.

Many diseases present with the same vague symptoms. For example, approximately nine respiratory diseases start with a cough, wheezing, droopiness, a pale head, and facial swelling. No friend, veterinarian, or poultry professional can diagnose over the phone or Internet (and to do so is fraught with possibilities of error and litigation). A personal examination of the bird, followed by detailed testing, is the best way to obtain an accurate diagnosis.

LIKE ALL OTHER BIRDS, CHICKENS NATURALLY SHUN OTHERS EXHIBITING SIGNS OF ILLNESS.

Examining your birds will make you accustomed to their individual qualities, making it easier to know when they may be ill.

Examining Your Birds

Getting to know your birds is the best assurance that you'll be aware when something's not right. Handle your birds frequently so you become accustomed to their sizes and weights. After doing so on a monthly basis, you'll have a better chance of detecting irregularities in the weight or overall health of your flock members.

If you detect illness, do a quick management assessment. Answer questions such as the following. Also, work with an extension agent to make sure you're making the right queries.

- ⤳ Have you recently switched feeds?
- ⤳ Have you introduced new birds recently?
- ⤳ Has the coop temperature recently dropped or increased?

Next, examine droppings in the coop or around the sick bird:

- ⤳ Do they look different than your flock's "normal"?
- ⤳ Is there blood in the droppings?
- ⤳ Do the droppings look particularly green (which means the bird isn't eating a balanced diet) or another color?
- ⤳ Are there undigested food pieces or worms in the feces?

You may notice that early in the morning, your birds' feces look green. That's normal. You're seeing bile from the gallbladder. Birds that have not eaten recently continue releasing bile into the digestive tract. It's not normal for a chicken's feces to stay green all day long for days at a time.

The feces of a bird not eating much may look green all day (versus the brown color of a bird eating normally). Check the mouth of a bird with green feces for an obstruction or lesion that may prevent normal feeding. If you are unsure whether a bird is eating a healthy amount, feel its crop at night, after the flock has gone to roost, to see whether it is full.

Use all five senses when examining your birds. Try these tricks:

*   **Examine sick birds for signs of injury.** Perhaps a disease agent is not to blame, but rather a bullying, dominant bird. Part the feathers to search for wounds. Watch the behavior at your flock.
*   **Listen to your birds' breathing patterns.** Any sign of wheezing, coughing, or gasping indicates respiratory distress. If a bird exhibits these symptoms, look for swelling around its eyes. Unlike humans, chickens have skin directly over their sinuses so an infection causes swelling of the face, particularly around the eyes. If the swelling becomes too severe, the bird can even have difficulty blinking.
*   **Check your birds' feathers.** Are they brittle or curled if they shouldn't be? These both indicate insufficient protein in the diet.
*   **Pay attention to muscle movement and balance.** Does the bird have difficulty maintaining balance? This can point to nervous system difficulties. If a bird limps, examine the bottom of its feet for signs of injury. In a wet litter, ammonia burns—black, scab-like blisters—can appear in the center of the foot or on the hock (the part of the leg where the feather meets the scale). You may also spot an injury called bumblefoot, which occurs when a sharp object pokes the foot pad and causes an infection in the foot.

Getting an Accurate Diagnosis

How can you advocate for your flock given all these potential dangers? Getting to know your birds is a great—and crucial—start. After that, you have more options than you think.

First, establish a relationship with a veterinarian who sees birds or exotics. This may mean driving an extra hour or two for an appointment, but consider it as an essential part of the challenge of raising poultry. If the only available option is your regular or livestock veterinarian, be aware that many will say off the bat that they do not work with poultry. Ask the vet to call a nearby poultry extension veterinarian, a program that offers phone consultation services for treating chickens.

Do your homework in advance so you have contact information ready for the nearest one. To find the appropriate extension veterinarian, call your local cooperative extension office (every county in the United States has one, for example, and universities in each Canadian province have associated agricultural extensions) or search for one online using terms such as *cooperative extension* and your county or town name.

Even with these resources, you may not get an instant diagnosis for your sick chicken. However, these specialists can recommend additional tests, the type of which you will help determine by your description of the chicken's symptoms. When this happens, your regular vet may need to send blood samples, fecal samples, or throat swabs to the nearest diagnostic laboratory. Some labs provide services free to owners of small flocks; others charge a nominal fee.

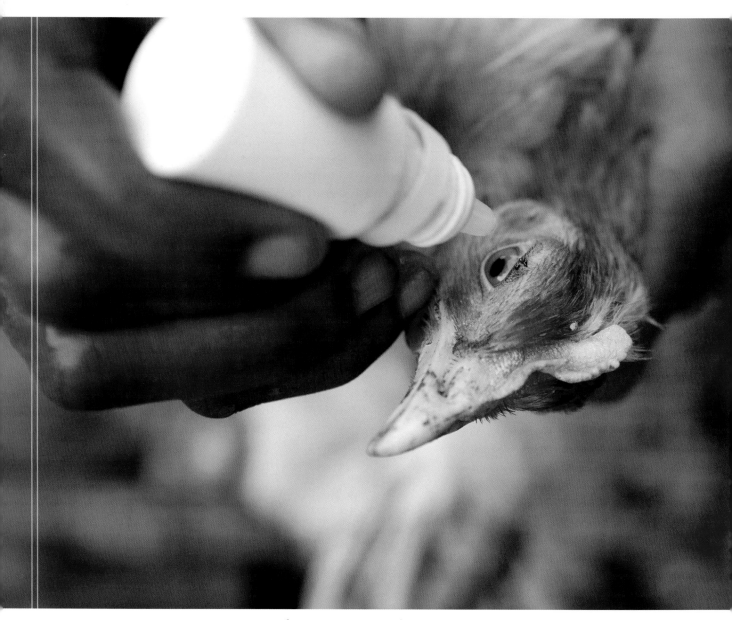

Administering an eye-drop vaccination

Vaccinations versus Antibiotics

With accurate diagnosis in hand, you can plan your attack on the disease agent. There is a difference between giving antibiotics and vaccinations. Vaccinations most often fight against viruses and with the exception of *Layrngotracheitis*, must be given prior to exposure to that virus. Just as in humans, this process allows the chicken to develop appropriate antibodies and the ability to fend off the virus.

Vaccinations are never a one-time deal. Just as humans require multiple doses of certain vaccines, so do chickens. To prime their immune systems, they receive mild forms of vaccines when they are young. Work with your vet to determine what you need to vaccinate for in your region of the country or the world. Once primed, the chickens get an annual shot of vaccine to ensure coverage for the year. Think of it like the annual flu vaccine for humans.

Antibiotics attack bacteria. They are short-lived and must remain in the body for several days to effectively fight the majority of invading bacteria. Antibiotics cannot defeat a virus and if given for an insufficient amount of time, may cause the bacteria to build up resistance rather than die. Once the immune system is down, any bacteria, including an antibiotic-resistant strain, can cause a serious secondary infection, which can lead to death. Throwing antibiotics at a flock wastes money and may cause antibiotic-resistant bacteria to proliferate. Confirm an accurate diagnosis before administering antibiotics to your flock.

Body Preservation

If a flock member dies, it is not only wise, but also prudent to send the body to a lab for a diagnosis. It may save the rest of your flock. Ignore the inclination to bury the chicken. Instead, triple bag the body, place it in the refrigerator, and treat it as evidence. Doing this presents the best chance of a lab making an accurate diagnosis. **Never place the bagged body in the freezer**; ice crystals can form inside body cells and destroy evidence pertinent for the lab's accurate diagnosis.

COMMON DISEASE ENEMIES

You've learned how to examine your flock and where to go for diagnoses. Now it's time to arm you with information about some enemies of a healthy flock. In the remainder of this chapter, we discuss several common poultry ailments, along with treatments and preventative measures. These are not the only poultry diseases. Be sure to seek an accurate diagnosis from a trained professional before beginning any treatment regimen.

Note: Many different diseases out there affect chickens and other poultry species. For example, one organism may be able to infect chickens and pheasants, while another can only cause illness in turkeys. Poultry germs come in the form of viruses, bacteria, fungi, and protozoa. Additionally, an arsenal of parasites out there can colonize a bird both externally and internally. For the purposes of this book, we focus on the most common diseases chickens can get.

Marek's Disease

Quick hit: Marek's disease is an incurable virus found worldwide that attacks a bird's nervous system, causing paresis or partial paralysis. Typical symptoms include poor motor control in the wings or legs on one side of the body. Often it is called range paralysis.

This incurable virus attacks a bird's nervous system, causing perisis or partial paralysis, such that at one time, it was known as range paralysis. This partial paralysis can be seen when one side of the body (wings or legs) suddenly exhibits poor motor control. Occasionally one leg will stretch out forward while the other goes backwards. The virus can also cause tumors and tremors, as well as immuno-suppression, opening up affected birds to infection by everyday germs. It typically strikes birds less than sixteen weeks old.

The virus spreads directly from bird to bird, but also through contaminated equipment, dust, bedding, down, and bird dander (the small flakes of feather follicle shed as a bird grows new feathers). It has an unusually long incubation period where a bird sheds virus and remains contagious for two weeks before showing any clinical signs of disease. The virus also resists many disinfectants.

The best prevention is to: vaccinate one-day-old chicks, no older. In commercial broilers or layer hatcheries, chicks are vaccinated inside the shell at day eighteen of embryo development to ensure that they hatch with antibodies to the virus. If buying your chicks at one of these hatcheries, order chicks already vaccinated against Marek's disease. If you choose to hatch eggs at home or let a setting hen do the work of an incubator, keep several bottles of Marek's disease vaccine on hand. Mix a fresh batch for each set of hatched chicks (the vaccine is only viable for a few hours after being mixed). Vaccinate chicks within their first twenty-four hours. Even just forty-eight hours post-hatch, their immune system has changed and they rapidly become less receptive to the vaccine.

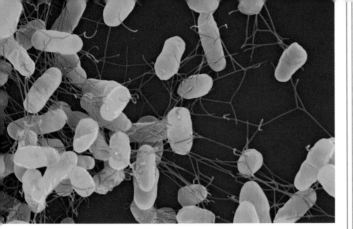

Salmonella

Salmonella

> **Quick hit:** There are more than 2,000 different types of *Salmonella*. Pullorum and fowl typhoid are two of the most common poultry diseases caused by *Salmonella*. Typical symptoms include diarrhea, high temperature, labored breathing, and a sicklike appearance.

The bacteria *Salmonella* is capable of colonizing many different animal species and environments. With that in mind, understand that they fall into two broad categories when it comes to poultry: *Salmonella* that colonize and cause illness in chickens only and *Salmonella* that colonize and cause illness in humans *through* chickens.

Many years ago, chicks being shipped post-hatch would die in transit or shortly thereafter due to the first type of *Salmonella*. This led to the creation of the (U.S.) National Poultry Improvement Plan (NPIP), which tests and then removes breeding adult birds that carry this form of the bacteria. If you plan to purchase chicks in the United States, always buy them from an NPIP-approved hatchery; these places take steps to ensure that chicks arrive at your door pullorum- and typhoid-free.

The second form of *Salmonella* infects humans through poultry and eggs, *Salmonella* Enteritidis (SE).

In the past several decades, SE has caused great concern due to multiple outbreaks in contaminated eggs. The hen lays eggs containing these bacteria. The bacteria then infect the human who eats the eggs raw or undercooked. That's why health experts always recommended fully cooking foods that contain eggs as an ingredient. If you're a flock owner who gives away, barters, or sells your extra eggs, keep in mind this recommendation about eggs being fully cooked before eating.

Your best bet to prevent infection by these bacteria—which can be present in your flock without birds showing any signs—is to use excellent biosecurity measures such as those we discussed earlier in this chapter and to fully cook your eggs to prevent food-borne illness. Probiotics—live organisms fed to the birds to outcompete bad bacteria—are an option for your poultry. If you find that your birds carry *Salmonella*, you may wish to give them antibiotics such as gentamicin and neomycin, two that have proven effective against the bacteria. Have a vet oversee dosages and duration of these antibiotics.

Mycoplasma

> **Quick hit:** Mycoplasma bacteria cause *Mycoplasma* Gallisepticum (MG), *Mycoplasma* Synoviae (MS), and *Mycoplasma* Meleagridis (MM), three common diseases in chickens, especially in backyard flocks. Typically, a flock member will exhibit cold-like symptoms of respiratory distress and general unthriftiness.

Mycoplasma take several weeks to grow in the lab before it's possible to make a definitive diagnosis. They are responsible for a series of respiratory symptoms known as chronic respiratory disease or CRD. The strains most often associated with illness in poultry are *Mycoplasma* Gallisepticum (MG), *Mycoplasma* Synoviae (MS), and *Mycoplasma* Meleagridis (MM).

Wherever you may purchase chicks, we recommend you ask hatcheries if they carry NPIP certification proving that their chicks are also free from MG, MS, and MM. Don't let the hatcheries tell you that *Mycoplasma* are problematic only in turkeys; this is not true.

Many owners of small flocks believe that it's normal for chickens to get a common cold in the fall, winter, or spring. It's not. Most frequently, cold-like symptoms result from an infection caused by *Mycoplasma* and poor biosecurity. *Mycoplasma* bacteria spread slowly through a flock, causing drops in egg production, an increase in thin-shelled or irregularly shaped eggs, or both. These infections are well known for their ability to strike laying hens with great frequency. There is a six- to twenty one-day incubation period for these bacteria to begin causing respiratory symptoms, as well as weight loss and swelling of the eyes. (Note: These symptoms also indicate several other respiratory diseases, so be sure to obtain a clear diagnosis.)

Prevention, made possible by solid biosecurity, is key with MG, MS, or MM. Vaccination also can prevent an outbreak. Purchase vaccine from manufacturers directly or in smaller quantities from veterinary supply companies. You can buy and administer this vaccine yourself. However, for a specific schedule, consult with a veterinarian. If you get a clear diagnosis that this organism is causing disease in your flock, give your birds the antibiotics Tylosin or LS 50 through their water.

Coccidiosis

> **Quick hit:** Common symptoms for a chicken affected by one of these parasites include a red or orange tint to feces, a drop in a chicken's feed consumption, and droopy or withdrawn behavior. Though this is an internal parasite, which we discuss in more detail below, we wanted to give it special attention because it's so common.

Coccidiosis or *Cocci* (pronounced *cock-see*) is an internal parasite of chickens. It is a protozoan rather than a virus or bacterium. Nine different coccidia infect chickens, though these coccidia cannot infect other animal species.

Chickens get coccidia through coprophagy, or eating feces. They eat their own, as well as that of other chickens, poultry, and wild birds. This is normal behavior. Once the coccidia get inside a chicken, they hatch, burrow into the intestinal lining, breed, and send eggs into the world through the bird's feces. If one chicken gets coccidia from infected feces, the parasite infects the ground of the coop. It can spread from bird to bird. Coccidia survive in the coop through damp ground, litter,

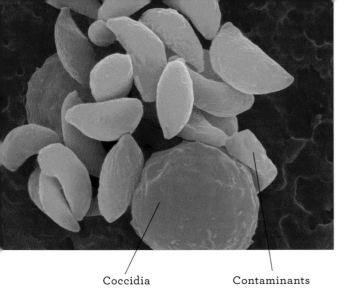

Coccidia   Contaminants

or unsanitary living/brooding conditions. This tends to hit chickens early, at about three to six weeks old, with the illness occurring most frequently at about one month of age. Chickens with coccidia may grow to adulthood and lay eggs, but their overall performance may never equal that of an uninfected bird.

The first symptom of *Coccidiosis* is a red or orange tint to the feces, either in lumps or in slick dropping from diarrhea. These colored lumps indicate damage to and bleeding from the gut lining as the coccidia burrow in. Soon bacteria gain entry to the bird's bloodstream and other tissues causing a secondary infection. Feed consumption drops and birds become droopy, withdrawn, and pale.

You have two choices at this point: medicate or let the disease run its course with the potential for some chickens to die. If you opt for the latter, birds that survive infection by one coccidia are not immune to the other strains. Before medicating, get an accurate diagnosis by having a regular vet perform a fecal flotation test on fresh droppings from an affected bird.

Once you identify that coccidia are to blame, work with your veterinarian to determine the most effective dosage for the drugs ALBON, Amprol, or Sulmet. Sulmet is the most commonly found at the feedstore, but it can damage the hen's reproductive tract. You cannot consume the eggs from a chicken that's received Sulmet because there is no withdrawal period.

Once again, prevention is key to keeping this disease out of your flock.

✦ Change bedding often, particularly once it gets damp.

✦ Do not overcrowd your chicks.

✦ Make sure you clean the brooder between sets of chicks. Your best bet is to raise your chicks on wire in a commercial-style brooder (see chapter 6 about brooders for a picture) to separate them from their feces.

You also can buy chicks from hatcheries that vaccinate for *Coccidiosis*. Contact your nearest veterinarian or extension agent for assistance if an outbreak of coccidia is confirmed.

Aspergillosis

Quick hit: Aspergillosis is a disease caused by a fungus. Typical symptoms include open-mouthed breathing or gasping for air, tremors, inability to balance, and head twisting. We wanted to give it special attention because it's so common.

Aspergillosis

Aspergillosis is caused by a fungus and can be especially damaging to chicks. Chicks have new cilia in their respiratory systems that do not operate perfectly until they are much older and therefore cannot fight off a rapidly moving fungal infection. Chicks show respiratory symptoms such as open-mouthed breathing or gasping for air. Also, infected chicks or adults may appear to have problems with their nervous system in the form of tremors, an inability to keep their balance, or even twisting their head around. This may be due to fungus growing in the brain or other tissues in the nervous system. Although this fungus can kill chicks, it may become a chronic problem in adult birds.

Airborne spores spread the fungus in warm, moist, and dirty environments. This occurs most often in an unclean incubator or brooder for chicks, but adult birds may become infected if the chicken coop is not kept clean. Eggs that come in to the incubator dirty may carry the spores from the fungus. Make sure that the eggs are laid in clean nest boxes. We do not recommend filling nest boxes with straw as it may carry the fungus.

Diagnosis usually comes by microscopic identification of the fungus found in infected internal organs such as the air sacs, lungs, or trachea. Symptoms of this disease look similar to Marek's disease. Regular cleaning reduces the presence of the fungus. Also, the addition of copper sulfate to the water helps prevent spreading the fungus through the flock. Drugs such as Nystatin and Amphotericin B are expensive but work to treat sick birds. Humans with suppressed immune systems can develop a similar form of pneumonia (mycotic pneumonia) caused by the fungus. Here's the fundamental message: Keep it clean!

Parasites

EXTERNAL PARASITES

External parasites most often come in the form of mites (chicken, northern fowl, and scaly-leg) and lice (chicken body and shaft). These parasites can actually suck the blood of the bird, irritating the animal day and night, preventing it from getting the restorative sleep it needs. This weakens the chicken's immune system, thereby opening up the bird to other diseases.

Check your birds weekly or monthly for signs of external parasites. To do so, part the chicken's feathers above and below the vent and look directly at the skin. Only scaly-leg mites are microscopic, so you will be able to see the others. Also look at the bird's feathers. Are they see-through or have lost their sheen? Do they have striations? These could indicate external parasites.

If your birds have external parasites, prepare for a long battle. You will not only need to treat the birds for several months, but you also will need to fully clean out the coop and spray it with the appropriate insecticide. Winter is when birds crowd together for warmth. Therefore, this is also when to increase your vigilance against external parasites.

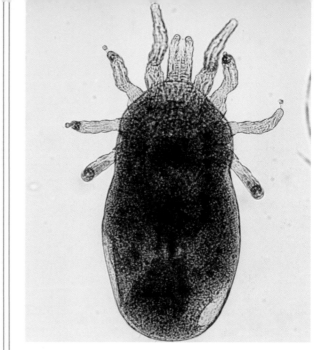

Chicken mite

Chicken Mites

Also called red mites, these parasites hide in the coop's nooks and crannies during the day and feed on birds at night. The chicken mite has been known to live in an environment for up to thirty-four weeks without anything to feed on. If they appear in your coop, make sure you thoroughly treat the chickens and clean the environment. To check for chicken mites, go to the coop after dark with a flashlight in hand. Lift the feathers on your birds' legs where the scales meet the feathers. Look for tiny red (after a meal of blood) or gray bugs. They do not move quickly and may even freeze at the sight of light.

Treat the birds and the entire coop to rid your flock of this problematic parasite. To start, wash the chickens weekly with a mild soap. You also can dust the birds with poultry dust (a chemical insecticide) or other carbaryl-based dust such as Sevin. This is time-consuming and you should make sure to wear a mask so you don't breathe in the dust. Other options are PERMECTRIN II 10 percent EC, solutions that come in concentrated form that you must dilute before spraying on chickens or in the coop. This type of insecticidal spray is quite effective and can be purchased at most feed stores. One bottle can last the owner of a small flock several years. Always mix these according to the directions on the label. And remember, treat the environment *and* the birds.

Northern Fowl Mites

These parasites are another type of mite that can harm your flock. They are brownish-gray and about the same size as the chicken mite, but spend their entire life cycle on the bird. Check your birds near their vents, particularly below the vent in the thicker feathers, for something that looks similar to dirt (but, of course, dirt doesn't move). Northern fowl mites also leave behind feces that look like the flea dirt seen on dogs or cats. Some other signs of this parasite are: whitish egg clusters at the base of the feathers, scarring, or red inflamed skin.

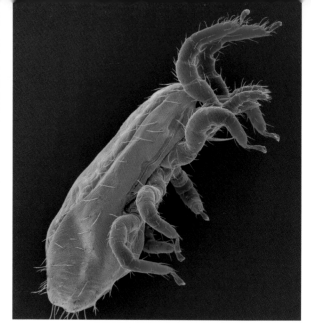

Northern fowl mite

This is a mite that in some cases may come into your coop via wild birds, so minimize your flock's contact with this potential hazard. If you spot northern fowl mites, use the same treatments and regimen for the birds and the coop as described for the chicken mite. Again, washing the chickens can work, but remember to treat the environment, too.

Chicken Body Louse

Chicken body lice are each half the size of a grain of rice and run very quickly. Their life cycle is short, so they reproduce quickly—meaning high numbers in short periods of time. They are light tan and most often appear around a chicken's vent. Part the feathers quickly and look down on the skin; the lice run for cover so be sure to check three or four places around the vent, under the wings, on the back, and on the breast. Also, look for clumps of whitish eggs glued to the feather fluff, very similar to the northern fowl mite.

Chicken body louse

Shaft louse

Chicken body lice can come to your flock via wild birds, so protect against this hazard. Louse can enter your flock by a newly introduced hen. Check for parasites during the quarantine period. Treat the birds twice in a ten-day period—once on day five and again on day ten—to ensure that you kill not only the adults, but also any young lice that hatched once the insecticide levels subsided. Once again, washing the chickens is a solution, but you also must clean the environment.

Shaft Louse

Shaft lice are harder to see. Darker in color and often found along a feather's main shaft, they tend not to move once the feathers are disturbed. They move slowly and lay eggs individually along the shaft of the feather. Sometimes on the primary and secondary wing feathers, you may see dark flecks that could be shaft lice. To check, pluck a feather (preferably not a primary or secondary wing feather) that contains a suspected louse and poke at the spot with a pin. After several prods, the louse may move on its own in an effort to relocate. Much

like the other mites and lice described, these come to your flock through contact with wild birds. Similar treatments apply.

Scaly-Leg Mite

This is a very different type of mite. It is microscopic, usually identified by the damage it does to any keratin-ized surface on a chicken (leg scales, shank scales, and beaks). The mite burrows through the surface to reach the skin's blood layer. It then multiplies, deforming the scales, which take on a flaky, crusty, irregular appearance. Soon large lumps appear on the legs; these can eventually cripple the bird. If left untreated, the beak may fall off or become permanently damaged.

To treat this ailment, suffocate the mites with an oil-based product. For at least two weeks, cover the chicken's legs daily in petroleum jelly, a camphor oil, or a half-kerosene, half-cooking oil mix. You may wish to wash the legs nightly in warm, soapy water to soften the scales. Gentle scraping also helps loosen and remove dead scales and mites. Continue this regimen until the legs return to their normal appearance.

INTERNAL PARASITES

These parasites come in the form of roundworms or tapeworms. There are several types of roundworms that can affect chickens including cecal worm and gapeworm. These breed in a chicken's gut and rob the bird of crucial nutrients. Roundworms can be treated using the one over-the-counter wormer on the market. Do not worm without verifying that your chickens actually have worms. Do this by looking at chicken feces for adult worms. You may not always be able to see adult worms, so take a sample of fresh feces to your nearest veterinarian to perform a fecal flotation test (as described in the *Coccidiosis* section). Roundworms eggs are small and can only be seen through a microscope. Adult roundworms are 1- to 1½-inches (2.5 to 4 cm) long and quite thin. To prevent reinfestation after treatment, clean out the bedding in the pen where the deworming treatment took place.

Tapeworms release their eggs in packets that can sometimes be seen moving in the feces. Once dried, these packets—about the size of a rice grain—open, spreading out the eggs. A tapeworm infestation can cause serious damage to the birds, feeding off their blood supply and stealing their nutrients, and unfortunately, they cannot be eliminated using the over-the-counter wormer described above. In fact, there are no over-the-counter tapeworm wormers suggested for use in poultry, so you need to seek a diagnosis, treatment, and prescription from a veterinarian.

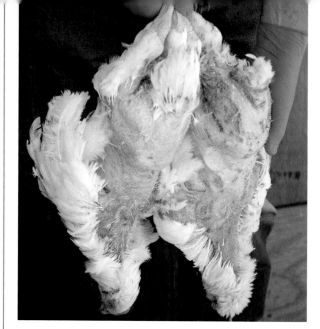

Poultry fallen victim to avian flu

Even if your birds don't exhibit symptoms of a worm infestation, for example, they're eating normally but losing weight, check their feces twice a year using the same fecal flotation test mentioned in the *Coccidiosis* section. This time, look for worm eggs. This is particularly important if you allow your birds access to the outdoors.

Avian Influenza and Newcastle Disease

You may have heard of the bad boys in the world of avian respiratory disease. Their names are avian influenza (AI) and Newcastle Disease (ND). Avian influenza made the news several years ago when an Asian strain of the virus began to infect humans. Symptoms of these viruses in chickens include coughing, sneezing, watery eyes, mucous, and any sort of respiratory symptoms.

AI and ND are game enders in the world of raising poultry. Detection of either disease typically causes a halt to the sale of poultry products from the country that is infected, resulting in economic distress for world poultry markets and poultry companies in that country. Keep your birds healthy and free of respiratory diseases, if not for the health of your home flock, then for the sake of all chicken owners.

If your chickens contracted a virulent form of AI or ND, the entire flock could be wiped out in two or three days. Any birds that lived would get tested and then tested again to rule out any false positives. At this point, someone from the government would likely have contacted you to ask you to avoid contact with other birds and to implement your most secure biosecurity measures. This is to keep whatever is on your farm on your farm and prevent it from infecting neighboring flocks. That means no birds wandering outside and no unnecessary travel for you to places where you could encounter other flock owners (for example, the vet's office, a feed store, a poultry club meeting, etc.). Consider yourself quarantined until you sort everything out.

If your birds come down with a virulent form of one of these diseases, they likely need to be put down. It's the humane way to act, as they are already suffering greatly. If dealing with a milder form of either disease, you unfortunately may still need to put down the birds as they can spread the virus to others or it can spontaneously become more virulent. If one of these diseases comes up, get help and fast so you can humanely euthanize flock members in serious distress.

Disease Reminder

A young, healthy flock, one not undergoing a molt, gladly lays eggs for you. Should the quality of eggs rapidly deteriorate or should egg laying stop altogether, take that as a sign that your flock is ill. Also, keep these tips in mind:

→ Take into account biosecurity as you build your coop. You hold the reins.

→ Don't think that a winter cold is the norm. It's typically a sign of something else going on.

→ Use the tools mentioned for getting a clear diagnosis before seeking a treatment.

→ Establish a relationship with your extension service and diagnostic lab. Ask them to work with you on creating workshops for the betterment of your flock.

→ Remember, treatments without a diagnosis are nothing more than wandering in the darkness of disease.

Spraddle leg and correcting spraddle leg at home

NONINFECTIOUS AILMENTS

You now know about infectious diseases that affect chickens. Let's turn to other noninfectious poultry health problems. These ailments, which typically surface from problems in management or housing, often get cured or can be prevented with simple changes to the coop or bird care.

Bumble Foot

Bumble foot is a foot infection that makes a bird limp or feel uncomfortable. It typically occurs because the bird has stepped on something sharp.

As soon as you see one of your chickens limping, look at its feet, specifically seeking out scars or scabs in the center of the three large toes. The center of the foot may also feel hot and appear swollen. If not caught early, this limp may become permanent, or infection could become systemic and the bird could die. If you can see the scab, one option is to use your fingers to push out and squeeze the puss and scar tissue building up in the wound. If you choose to perform this procedure on the bird, wash the wound, wrap the foot in gauze immediately, and then

during recovery, place the bird in a sick pen that contains a great deal of deep bedding. Rinse the bird's wound daily with a warm soap that contains iodine, apply a wound spray, and then rewrap the foot. Also pay attention to the sick bed, cleaning it daily to remove feces or wet bedding. You'll know the bird has recovered when it starts walking normally again. Some physical therapy will help keep both legs strong during recovery.

Remember to look for the source of the injury, so it won't harm other chickens. Touch all surfaces of the coop with your bare hands, feeling for something sharp. Remove the culprit or make repairs where necessary.

Spraddle Leg

Spraddle leg, which typically occurs during brooding, is just as it sounds: A slick surface causes a chick's legs to spread out in opposite directions, to the left and right, causing slipped tendons. The condition is painful for the chick and can cause permanent damage to the legs, eventually preventing the chick from being able to walk. Sadly, the baby bird will eventually starve and die because it cannot reach food or water.

Curled toes and correcting curled toes at home

You can fix spraddle leg if you identify it early. Create a small splint using a pipe cleaner or plastic bandages. If using a pipe cleaner, cut off a 3-inch (8 cm) piece and gently wrap one end around one leg. Lightly pull the other leg into a normal position, and then wrap the other pipe cleaner end around the other leg. Most chick owners splint the chick themselves rather than going to a vet.

To prevent spraddle leg, avoid using newspaper or magazines in your brooder as flooring; they are too slick. Instead try wood shavings, wire, or even a nubby old terrycloth towel. (Look back to chapter 6 for a refresher on brooders.)

Curled Toes

Newly hatched chicks with this condition cannot unfurl their toes and therefore cannot walk normally. This can occur because of poor genetics or the breeding adults' diet. It also can come from errors in temperature or humidity during incubation. If not corrected, this can lead to sores on the chick's hock and breast and eventually starvation because the chick can't walk properly.

Fix the problem by uncurling the toes and taping them into the correct position. Use a small piece of heavy paper and tiny strips of masking tape to arrange the toes into the correct position. You can also try plastic bandages. Uncurl the toes, place them in the correct positions in the bandage's center, and then wrap the adhesive ends over the top of the toes to hold them in place. Be sure to trim any excess material. You'll need to change these dressings daily because chicks grow rapidly—even in a few days' time.

Chickens can develop frostbite on their extremities if exposed to extreme cold temperatures for an extended period.

Frostbite

Frostbite often occurs on a bird's comb, wattles, and toes, potentially causing these features to turn yellow or black. This dead tissue may fall off over time.

Breeds with large, pendulous wattles get these appendages wet every time they drink. Also, the comb of some chickens becomes so large that it's unable to fit entirely under the bird's wing at night. And incorrectly sized roosting poles—too small or too large in diameter—can prevent the birds from pulling their breast feathers over their toes at night for warmth.

Solve this winter problem by keeping the coop at temperatures above freezing. Also, use a heated water base to keep the chickens' water above freezing. Finally, at night before the birds roost, try spreading a small amount of petroleum jelly over the combs and wattles of birds with large combs. However, note that this gets messy quickly because the jelly spreads on to the feathers. Messy feathers pick up dirt, so you'll need to wash these birds periodically. If temperatures dip too low, even the petroleum jelly may freeze, so your best bet is a warm coop.

Egg Bound

Egg-bound hens experience difficulty passing an egg normally. Whether a chronic problem or a singular event, being egg bound is a dire situation for the hen. These birds take on a penguinlike stance and their abdomens get hot. They also appear strained. Depending on the hen and the situation, the egg may eventually pass. However, sometimes the egg gets caught in the hen permanently, which may eventually result in death. Also, if the egg breaks inside the hen, she will die.

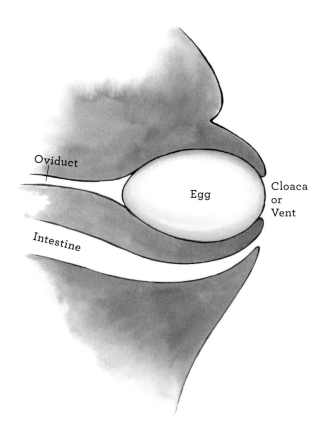

Oviduct

Egg

Cloaca
or
Vent

Intestine

Internal Layer

An internal layer is a hen that passes her yolks not into the oviduct (the correct place), but rather into her abdominal cavity. Older hens with less efficient reproductive tracts may become internal layers. You can tell this is going on if the abdomen becomes hot or distended, indicating an abdominal infection. Sadly, there is nothing that you can do for these hens as they often pass yolks into their abdominal cavities daily. Most internal laying hens die.

Prolapsed Vent

A prolapsed vent—when the oviduct has been pushed out of the body as the hen tries to lay eggs—often occurs in young birds. The eggs may be too large for the not-fully-developed oviduct. In other words, if a hen is exposed to too much light before her reproductive tract matures, she will try to lay eggs before her body is completely ready. Light-bodied hens such as Leghorns should start laying at about sixteen to eighteen weeks. Heavy hens, such as most hybrid brown-egg layers, begin around eighteen to twenty weeks.

Try to help the hen by immersing her vent and backside in warm water, then very gently massage the area around the egg. Be gentle when handling the hen's abdomen and careful around the egg or it may break. Another option entails putting a small amount of mineral oil on a latex glove and gently massaging the oil up the chicken's vent and into her oviduct to aid the egg's exit. If the hen's quality of life is no longer sufficient, you may consider culling her.

Never breed egg-bound hens because the problem can be genetic.

If you find a hen with a prolapsed oviduct, separate her from the rest of the flock immediately. The oviduct will be shiny and attractive to other chickens, which may come over and peck at it. This type of cannabilism can quickly kill the affected hen. Thoroughly wash off the prolapsed oviduct and then wash your own hands. Using a glove and mineral oil, push the oviduct back in through the vent. Once a hen prolapses, she is prone to doing so again, so keep her separate in the sick pen for several days as her body matures and heals.

APPENDIX

There is a variety of colors and sizes of eggs that can add another layer of interest to some of the artistry that can be created with them.

EGG ARTISTRY: WHAT TO DO WITH ALL THOSE EGGS (BESIDES EAT THEM)

Now that you've begun raising chickens in your backyard, you'll soon have more eggs than you know what to do with. Eggs aren't just for eating. Here are some crafty, fun ways to use your eggs. You may be familiar with a decorative piece that sits on an egg stand or strung as a hanging ornament. But there is a world of variation within this art.

Begin your egg art with almost any kind of egg—duck, goose, turkey, quail, emu, or even ostrich, to name a few—but we focus on the chicken egg because it's the simplest and you likely have an abundance. (If you're interested in purchasing other types of eggs, try www.theeggeryplace.com or www.metzerfarms.com.)

Blowing Out an Eggshell

To use an eggshell as your canvas, often you need to blow out the shell's contents first. Poking a hole at either end to do this is a simple method, but there are better ways, especially because this two-hole method requires your lips to make direct contact with the eggshell. (If you opt for this method, be sure the egg is clean and avoid consuming its contents due to the risk of food-borne illness such as *Salmonella*.)

Marans produce richly colored brown eggs.

Here's another, more preferable way to blow out a chicken egg: the one-hole syringe method.

① Purchase a syringe with an 18- or 20-gauge 1½-inch (4 cm) needle.

② Make three holes very close together in a triangle shape, within the same ⅛ of an inch (⅓ of a cm).

③ Create one larger hole by connecting the three you made initially.

④ Stick the needle into the egg as far as it will go and wiggle it around until you poke a hole in the yolk. Attach your needle to a 10 cc syringe. (This is standard syringe language and how you would ask for it.)

⑤ Fill the syringe with air and invert the egg so the hole faces down. Place the needle attached to the syringe into the egg and slowly push up on the plunger. Air enters the egg and the egg's contents exit through the hole.

⑥ Continue injecting air until all of the contents have been removed. The contents of the egg are still edible, so feel free to place the yolk and egg white into a bowl, cover it, and refrigerate it.

Rinsing the Shell

Pour warm water into a bowl and add a dash of liquid hand soap or dish detergent.

① Fill your syringe with the water mixture and inject it into the empty shell. Repeat until the eggshell fills with soapy water.

② Cover the hole with a finger and shake the egg to coat all inside surfaces with the soapy solution.

③ Wait two minutes and then blow out the soapy water using the same method you used to blow out the yolk and egg white.

④ Rinse out the soapy suds using warm or cold water. Repeat until no more soap or bubbles come out of the egg.

⑤ Let the egg drip dry, hole-side down, on a paper towel for twenty-four hours.

Once dry, your eggs are ready for decorating.

Fabergé Eggs

Peter Carl Fabergé, one of the most famous egg artists, was a jeweler for the House of Fabergé. Between 1885 and 1917, he created bejeweled eggs for Russian tsars and wealthy Moscow families. The eggs ranged in size from necklaces to large, tabletop decorations. Almost all were adorned with precious metals and jewels in intricate arrangements. At one time, sixty-five of these eggs existed, but today, only fifty-seven are known to have survived.

The popularity of Fabergé eggs spurred an interest in a new field: handcrafted, decorated or jeweled eggs. Today, many businesses sell specialty hinges, filigree work, and clasps for delicate egg crafting. If this sparks your curiosity, check out www.alcrafteggartistry.com, www.eggersdelight.com, or www.eggsbybyrd.com.

Japanese Washi Eggs

Japanese washi eggs are blown out and wrapped in washi paper, which is made of rice. The designs typically mimic scenes and patterns found on the fabrics used to create kimonos. The size of the egg doesn't matter as long as the paper covers the egg.

Pysanky Eggs

Pysanky is a traditional Ukrainian method of egg decorating. Using a wax-resist method and strong dyes, eggs are decorated in complex geometrical shapes and patterns. Each symbol and color has a different meaning. A deer or stag, for example, symbolizes strength. A bushel of wheat symbolizes

Fabergé eggs

Washi eggs

Pysanky eggs

a good harvest. There are also two styles of pysanky: traditional and trypillian. The former uses geometric shapes and bright colors and lines; the latter uses more swirls and earth tones.

Pysanky eggs are decorated as follows:

Beeswax is melted and funneled down to a point using a device called a kitska. The point is then used to draw onto an egg that still contains its egg white and yolk (a blown-out one will not sink down into the dye). The egg is then dipped into a dye, removed, and dried. Any area the artist wishes to keep a specific color gets covered with wax before the next round of dye. The final dye is usually the darkest color, often black.

To remove the beeswax from the egg, place the egg on a wooden board containing three nails in triangular position. Put the board and egg into an oven set to 200°F (93°C). This will cause the wax to melt, allowing you to gently wipe it off with a paper towel. Once the egg dries, the wax melts off to reveal meticulous patterns and symbols.

You can purchase a dye kit for starting your own pysanky egg art at stores such as the Ukrainian Gift Shop at www. ukrainiangiftshop.com.

Czechoslovakian Drop-Pull Eggs

This decorative technique is similar to pysanky. It also uses melted wax, but instead of beeswax alone, the mixture is 50 percent beeswax and 50 percent paraffin wax with a little wax dye for color. (You can find wax dyes and paraffin wax in the candle-making section of your nearest craft store. Purchase beeswax from a local beekeeper.)

Using a kitska to make a pysanky egg

Drop-pull eggs are decorated as follows:

The waxes and dyes get melted together and stirred until smooth. A pattern is drawn onto an egg using pencil. With the design in place, the round end of a sewing pin stuck into the eraser end of a pencil gets dipped into the molten, dyed wax to allow it to heat up. Very quickly, it's pulled out and placed onto the egg, leaving a dot of colored wax. Then comes the pull, by making a short stroke, starting at the dot. Start with a painted egg for added interest.

Decoupage

Decoupage entails gluing cutout images onto an empty blown-out eggshell and lacquering them into place. Quick and easy, this craft is good for beginners. You can paint the egg before applying the pictures.

Carved Eggs

Delicate and beautiful, carving eggs takes patience, practice, and many a broken shell to master. Intricate designs include not only holes but removal of layers revealing hidden colors and surprises. An emu's egg, for example, is a dark blue-green color that changes to teal the lower you carve into the eggshell. Beneath the teal is a white eggshell. If you're interested in taking up this form of egg artistry, visit www.theeggshellsculptor.com to learn more from an egg sculptor.

Scratched Eggs

Eggs decorated with the scratch method begin with an egg that is not blown out. Start with a painted or dyed egg. (The coloring on dyed eggs doesn't tend to flake off as can sometimes happen with a painted egg surface.) Using a craft knife or a scalpel, scrape away the design you want to create. Once complete, blow out the contents of the egg and spray the outside with a clear acrylic for shine.

Decoupaged egg

Carved eggs

Scratched eggs

RESOURCES

GLOSSARY

Bantam: This is the smaller counterpart to a large fowl breed. Every large fowl has a bantam option too. Some breeds come only in bantam size and are therefore called true bantams.

Bloom: A protective coating on a just-laid egg

Brood patch: An area on the breast with no feathers and increased vascularization where the eggs come in direct contact with the hen's skin during incubation

Brooder: This is a box or pen that provides a warm, dry home for new chicks during their first weeks of life. It contains a fresh and continuous supply of food and water and protection from predators.

Broody: This is a hormone-driven condition that occurs when a hen is willing to sit on and hatch eggs. It can become detrimental to the hen if she will not conduct other regular activities such as eating or drinking.

Chicken: A domesticated fowl, originally from Southeast Asia, used for both meat and eggs

Cocks: Male chickens older than one year of age, sometimes called roosters

Cockerels: These are male chickens younger than one year of age. Females are called pullets.

Crop: The outpocketing for the esophagus in the neck region just above a chicken's breast

Crumble: A feed mixture for chickens that includes broken up pellets in small pieces

Dander: The small flakes of feather follicles shed as a bird grows new feathers

Egg tooth: This is the hard part of the beak a chick uses to break through its shell during hatching. It falls off a few days after the chick hatches.

Feeder: The container in both the brooder and the coop where food is stored

Fluff: A chick's first feathers, technically called natal down

Hen: Female chickens older than one year of age

Hock: The joint of a chicken's leg where the feather meets the scale

Marek's disease: A virus found worldwide that attacks a chicken's nervous system, causing partial paralysis.

Mash: Food for chickens that's a powdery mixture of chicken feed

Natal down: See fluff.

Oviduct: The organ in the hen which accepts the yolk after ovulation, where the egg is completed

Pasty vent: A collection of fecal matter around the chick's vent that a hen would take care of in nature but that a chick owner must clean

Pipping: This is the two-part process by which chicks hatch. During internal pipping, the chick breaks through the air cell inside the egg and takes its first breath. The second stage is external pipping, during which the chick makes its first break through the eggshell.

Primary wing feathers: These are part of the wing. When you fold out the wing, the primary feathers are furthest from the body.

Pullets: These are female chickens younger than one year of age.

Roosting pole: The place where chickens perch and sleep at night

Run: The fenced-in area outside the coop where chickens can range, scratch for bugs, and take dust baths

Secondary wing feathers: These are the wing feathers closest to the body. When a chicken folds up its wing, secondary feathers are visible. Primary feathers are tucked up underneath.

Waterer: The container in both the brooder and the coop where water is stored

Vent: The opening through which a hen lays her eggs or any bird drops its fecal matter.

WEB RESOURCES

ALCRAFT EGG ARTISTRY, LLC
www.alcrafteggartistry.com

AMERICAN BANTAM ASSOCIATION
www.bantamclub.com

AMERICAN LIVESTOCK BREEDS CONSERVANCY
www.albc-usa.org

AMERICAN POULTRY ASSOCIATION
www.amerpoultryassn.com/breed_classifications.htm

BACKYARD CHICKENS
www.backyardchickens.com

BRAIDED BOWER FARM
www.braidedbowerfarm.com

CHICKEN WHISPERER, INC.
www.chickenwhisperer.net

THE CITY CHICKEN
www.thecitychicken.com

ECKERSLEY'S ART & CRAFT
www.eckersleys.com.au

EGGS BY BYRD
www.eggsbybyrd.com

EGGERS DELIGHT
www.eggersdelight.com

THE EGGSHELL SCULPTOR
www.theeggshellsculptor.com

THE EGGERY PLACE
www.theeggeryplace.com

FEATHER SITE
www.feathersite.com

INTERNATIONAL EGG ART GUILD
www.eggartguild.org

INTERNET CENTER FOR WILDLIFE
DAMAGE MANAGEMENT
www.icwdm.org

METZER FARMS
www.metzerfarms.com

MUNICIPAL CODE CORPORATION
www.municode.com

MY PET CHICKEN
www.mypetchicken.com

www.Poultrymad.co.uk

http://poultrybookstore.blogspot.com

SOCIETY FOR PRESERVATION
OF POULTRY ANTIQUITIES
http://poultryb.dot5hosting.com/sppapage.html

TRACTOR SUPPLY
www.tractorsupply.com

UKRAINIAN GIFT SHOP
www.ukrainiangiftshop.com

UNIQUELY EMU PRODUCTS, INC.
www.uniquelyemu.com

URBAN CHICKENS
www.urbanchickens.org

YAHOO GROUPS
//tech.groups.yahoo.com/group/CHICKEN-101

//tech.groups.yahoo.com/group/OrganicChickens

MAIL-ORDER POULTRY SUPPLIES

BOWLES POULTRY SUPPLIES
Lucasville, Ohio
(740) 372-3973

BRINSEA
www.brinsea.com

CHARLIE'S POULTRY SUPPLIES
For-Valley, Virginia
(540) 933-6123

CLAUSING COMPANY
Nocatee, Florida
clausing@desoto.net

CRAZY K FARM
www.crazykfarm.com

CRITTER CAGES.COM
www.critter-cages.com

CUTLER'S PHEASANT, POULTRY, AND
BEE SUPPLIES
www.cutlersupply.com

EGGANIC INDUSTRIES
(800) 783-6344
www.henspa.com

EGGBOXES.COM
(800) 326-6667
www.eggboxes.com

FLEMING OUTDOORS
(800) 624-4493
www.flemingoutdoors.com

FIRST STATE VETERINARY SUPPLY
www.firststatevetsupply.com

MAIL-ORDER POULTRY SUPPLIES *continued*

GQF MANUFACTURING COMPANY INC.
www.gqfmfg.com

IDEAL POULTRY BREEDING FARMS, INC.
www.ideal-poultry.com

KEMP'S KOOPS
www.poultrysupply.com

MURRAY MCMURRAY HATCHERY
www.mcmurrayhatchery.com

MT. HEALTHY HATCHERIES INC.
www.mthealthy.com

AMISH GOODS
www.myamishgoods.com

MY PET CHICKEN
www.mypetchicken.com

BIDBIRD AUCTION SITES
www.bidbird.com

POULTRYMAN'S SUPPLY COMPANY
www.poultrymansupply.com

RANDALL BURKEY COMPANY, INC.
www.randallburkey.com

SEVEN OAKS GAME FARM
www.poultrystuff.com

SHOP THE COOP
www.shopthecoop.com

SMITH POULTRY AND GAME BIRD SUPPLIES
www.poultrysupplies.com

ORGANIZATIONS

AMERICAN LIVESTOCK BREEDS CONSERVANCY
www.albc-usa.org

APA–ABA YOUTH PROGRAM
Philadelphia, Tennessee
nanamamabrahma@att.net
www.apa-abayouthpoultryclub.org

AVEC
Association of Poultry Processors, Europe
www.avec-poultry.eu

FFA
www.ffa.org

4-H
www.4-h.org

PUBLICATIONS

4-H GUIDE: RAISING CHICKENS
Tara Kindschi

AMERICAN STANDARD OF PERFECTION
Various Artists

THE CHICKEN HEALTH HANDBOOK
Gail Damerow

BACKYARD POULTRY
www.backyardpoultrymag.com

THE CHICKEN HEALTH HANDBOOK
Gail Damerow

CITY CHICKS
Patricia Foreman

HOW TO RAISE POULTRY
Christine Heinrichs

POULTRY PRESS
www.poultrypress.com

RAISING POULTRY THE MODERN WAY
Leonard S. Mercia

*YOUR CHICKENS: A KID'S GUIDE TO RAISING
AND SHOWING*
Gail Damerow

INDEX

PHOTOGRAPHER CREDITS

© ABN Images/Alamy, 63 (bottom, right)
AFP/gettyimages.com, 49
© Ambient Images Inc./Alamy, 9
Peter Anderson/gettyimages.com, 55 (right)
© Andia/Alamy, 54
© ARCO/De Meester/agefotostock.com, 48
© Arco Images GmbH/Alamy, 119 (left)
© ARCO/Reinhard, H./agefotostock.com, 123
© Art Directors & TRIP/Alamy, 124
Rick Bennett, 96 (right)
Courtesy of Jeannette Beranger/The American Livestock Breeds Conservancy/www.albc-usa.org, 25; 26, 28 (right); 29; 30; 31; 32; 33; 34 (right); 38 (left); 41; 135
Michael Blackwell/www.michaelblackwell.com, 161
© blickwinkel/Alamy, 63 (top, middle)
© Bill Boch/agefotostock.com, 114
© Bon Appetit/Alamy, 126 (right)
Tom Brakefield/gettyimages.com, 119 (right)
Courtesy of Corallina Breuer, 5; 42 (left)
© Paul Cannon/Alamy, 118 (left)
Carolina Biological Supply Co./Visuals Unlimited, Inc., 151
© Corbis Premium RF/Alamy, 28 (left)
© Sylvia Cordaiy/Alamy, 138
© Johan De Meester/agefotostock.com, 60
Eckersley's Art & Craft/www.eckersleys.com.au, 162 (middle)
© FLPA/Alamy, 95
© Focus Database/agefotostock.com, 50
Fotolia.com, 14 (right); 59; 61; 62 (top, second & third, bottom, left; third row, right; bottom row); 63 (top, left & right; second row; bottom, left); 77 (bottom, left); 101; 104; 116 (left & right); 117 (left); 118 (right); 121 (bottom); 129; 130; 132; 164 (top & bottom)
© Stephen French/Alamy, 109
© geogphotos/Alamy, 126 (left)
© Les Gibbon/Alamy, 10
© Rob Gibbons/Alamy, 131 (third row, middle)
Courtesy of GQF Manufacturing, 68; 70 (bottom); 72; 73 (top)
© Green Stock Media/Alamy, 12; 27 (left)
© Grant Heilman Photography/agefotostock.com, 107
© Grant Heilman Photography/Alamy, 125 (right)
© Image Source/Alamy, 65

iStockphoto.com, 2; 3; 7 (left); 11 (top); 17; 34 (left); 39; 40 (right); 42 (right); 43; 44; 57; 58; 67; 70 (top); 73 (bottom); 77 (top); 88; 89; 90; 92; 116 (middle); 117 (left); 120; 121 (top); 160; 162 (top & bottom); 163; 100
© E. A. Janes/agefotostock.com, 77 (bottom, right)
© Ernie Janes/Alamy, 35
© Nick Jenkins/Alamy, 47
Andrea Jones/Photolibrary, 15
Kim Karpeles/Photolibrary.com, 100
Michael Kelley/gettyimages.com, 53
© Don Kohlbauer/zumapress.com
kpzfoto/Alamy, 97
Dennis Kunkel Microscopy, Inc./Visuals Unlimited, Inc., 152; 153 (left)
© Jake Lyell/Alamy, 144
Courtesy of Mike MacDonald/mrfancyplantsnursery.com, 37
© J. Marshall/Tribaleye Images/Alamy, 113
© Mauritius images GmbH/Alamy, 136
© mikecranephotography.com/Alamy, 102
Courtesy of Teri Myers & Peter M. Bergin/braidedbowerfarm.com, 6 (left); 7 (right); 74-75
National Geographic/Getty Images, 164 (middle)
© National Geographic Image Collection/Alamy, 36; 56
© The National Trust Photolibrary/Alamy, 38 (right)
© Ulrich Niehoff/agefotostock.com, 131
© B. O'Kane/Alamy, 19
© Micheline Pelletier/Sygma/Corbis, 40 (left)
© PHOTOTAKE Inc./Alamy, 147; 149; 150
© Picture Contact BV/Alamy, 125 (left)
© Simon Price/Alamy, 158
© redbrickstock.com/Alamy, 24
© RIA Novosti/Alamy, 137
© Valery Rizzo/Alamy, 140
Andy Schneider, 79; 81; 82; 84; 86; 91; 93
Science Source/Science Photo Library, 71
Rosalind Simon/Photolibrary.com, 96 (left)
Zia Soleil/gettyimages.com, 110
© Inga Spence/Alamy, 142; 154
© Rob Walls/Alamy, 14 (left)
© Peter Webb/Alamy, 11 (bottom)
William Weber/Visuals Unlimited, Inc., 27 (right)
© Greg Wright/Alamy, 55 (left)

ABOUT THE AUTHORS

Andy Schneider, better known as the Chicken Whisperer, has become the go-to guy for anything chickens. Over the years, he has helped a countless number of people start their very own backyard flocks. He is not only a well-known radio personality, as host of the "Backyard Poultry with the Chicken Whisperer" radio show, but also a contributor for *Mother Earth News, Grit* magazine, and *Farmers' Almanac*. He is the national spokesperson for the USDA–APHIS Biosecurity for Birds Program. He is the founder and organizer of the Atlanta Backyard Poultry Meetup Group and many other meetup groups around the United States. He has been featured on CNN, HLN, Fox, ABC, CBS, NBC, NPR, as well as in the *Wall Street Journal, Time, The Economist, USA Today, New Life Journal*, and other national and local publications. More recently, Andy has been traveling around the U. S. on the Chicken Whisperer Tour, educating people about the many benefits of keeping a small backyard flock of chickens. Andy and his wife Jen keep 35 chickens on their property just north of Atlanta, Georgia. See his website: www.chickenwhisperer.net.

Dr. Brigid McCrea, Ph.D., began her lifelong love of chickens quite by accident, and she attributes it all to her involvement in 4-H. With the help of the family mechanic, who was also a show chicken breeder and a 4-H poultry leader, she began to raise and show chickens. The poultry community is full of kind people willing to share their knowledge with fellow poultry enthusiasts, and it was this camaraderie that spurred her on to study poultry and birds in college. She received her B.S. and M.S. in avian sciences from the University of California, Davis, and then received her Ph.D. in poultry science from Auburn University. Dr. McCrea is pleased to share her poultry knowledge with everyone who is interested in learning how best to start or serve their flock.

ACKNOWLEDGMENTS

First and foremost, I would like to thank almighty God for the many blessings he has given me, saving my soul, forgiving me for my sins, and giving me eternal life. I would like to thank my wonderful parents, Guy and Shirley, for teaching me that anything is possible if I put my heart and soul into it. I would like to thank my beautiful wife, Jen, for all of her loving support. She is not only my wife, but also my best friend. Through her love, compassion, forgiveness, and understanding, she has not only made me a better person, but a better Christian, and I look forward to spending the rest of my life with her. Finally, I would like to thank all of my friends and fans for encouraging me to continue doing what I love to do, and supporting me along the way.

Dr. Brigid McCrea would like to acknowledge Andy Schneider for asking her to participate in the development of this book. His unwavering enthusiasm for backyard poultry is loved by all of his radio listeners and guest speakers alike. Dr. McCrea would also like to thank all of her poultry extension mentors at the University of California, Davis, and at Auburn University: Dr. Francine Bradley, Dr. Joan Schrader, Dr. Sarge Bilgili, Dr. Joseph Hess, and Dr. Kenneth Macklin. If not for these individuals and their enthusiasm for poultry science, she would never have chosen the career that she did.